Volume 93   Number 3   September 2021

American Literature

The Infrastructure of Emergency

*Edited by Stephanie Foote, John Levi Barnard,
Jessica Hurley, and Jeffrey Insko*

Essays

Book Reviews

Jessica
Hurley
and
Jeffrey
Insko

Introduction:
The Infrastructure of Emergency

On July 16, 1979, the largest radiological disaster
in United States history took place in New Mexico when the failure of
a tailings dam at the United Nuclear Corporation's Church Rock uranium mill led to the release of 1,100 tons of radioactive mill waste and
95 million gallons of highly acidic, highly radioactive liquid effluent
into Pipeline Arroyo, from where it entered the Río Puerco. Following
its course though the Navajo Nation, the irradiated river left radio-
toxic sediments and radioactive groundwater in wells and aquifers
across Dinétah. Built on land known to be geologically unsound and
displaying large cracks as early as 1977, the dam was known by both
the United Nuclear Corporation (UNC) and the state and federal
agencies that had granted its construction license to be an unstable
infrastructure on shaky ground (Brugge, deLemos, and Bui 2011).
But this was Navajo ground, and the violence was slow, and the mill
produced $200,000 in yellowcake per day, and so the risk of catastro-
phe was ignored until it was actualized—at which point it was essentially ignored again, overshadowed by the Three Mile Island release
that had occurred four months earlier, which had been more spectacular and impacted mostly white settlers rather than Diné. Desultory
cleanup efforts by first the UNC and then the Environmental Protection Agency (EPA) have left the area widely contaminated; as of January 2021, "groundwater migration is not under control" (EPA n.d.).[1]
The devastating health effects of long-term exposure to radiotoxins
continue to impact the Navajo Nation, where they both compound and
are compounded by the social and bodily harms of life lived under
colonial occupation (Voyles 2015: 4).

Thirty years later and 1,500 miles away, the most expensive inland

American Literature, Volume 93, Number 3, September 2021
DOI 10.1215/00029831-9361209 © 2021 by Duke University Press

oil spill in US history took place near the city of Marshall, Michigan. On July 25, 2010, an Enbridge oil pipeline transporting diluted bitumen from the Alberta tar sands ruptured, spilling more than a million gallons of oil into Talmadge Creek, a tributary of the Kalamazoo River. Oil gushed from a six-foot-long gash in the thirty-inch steel pipeline for seventeen hours before the leak was finally detected. Meanwhile, the oil reached the Kalamazoo, where it traveled another thirty miles downriver, saturating the riverbank and coating flora and fauna. Compounding the spill and confounding first responders, the oil that spilled did not float on the surface of the water. Instead, the chemicals used to dilute the thick crude for pipeline transport, which included the known-carcinogen benzene, evaporated into the air. The remaining bituminous material then sank to the bottom of the river, rendering ineffectual ordinary containment efforts. Years before the spill, safety inspections on the pipeline had revealed serious defects, but weak federal oversight and a corporation focused on profits rather than safety disincentivized mitigating action (National Transportation Safety Board 2012).[2] Despite its significance as the first major release of tar-sands oil into a body of freshwater, the Kalamazoo River spill attracted little national attention, overshadowed by the more spectacular, and deadly, Deepwater Horizon blowout in the Gulf of Mexico just a few months prior. After a seven-year, billion-dollar cleanup effort, thousands of gallons of oil remain in the Kalamazoo's riverbed and the public health effects upon local residents exposed to contaminated air remain entirely unstudied, despite anecdotal accounts of a variety of illnesses.

Turning our attention to Church Rock and Marshall rather than Three Mile Island and Deepwater Horizon, we mean to highlight not the sensational event but the routine character of infrastructural violence, not the singularity of disaster but its ongoingness under conditions of colonial racial capitalism. The Río Puerco and the Kalamazoo River, for example, had been polluted long before the collapse of the UNC tailings dam and the rupture of the Enbridge pipeline. From 1968, when the Río Puerco became a perennial stream, to the mid-1970s, there was no regulation of water quality downstream from the mines; as Traci Brynne Voyles (2015: 165) notes, "New Mexico Environmental Improvement Division (EID) reports indicated that the river 'contains levels of radioactivity and certain toxic metals that approach or exceed standards or guidelines designed to protect the health of people, livestock and agricultural crops,' which were its primary uses by Diné residents of the Church Rock area." Similarly, the

Kalamazoo River had suffered various forms of pollution going back to the nineteenth century; one researcher in the 1960s famously described it as having the color and consistency of a blueberry milkshake (Dempscy 2001: 1). In 1990, the US EPA declared portions of the river a superfund site, owing to high levels of polychlorinated biphenyls (PCBs) from toxic waste dumped into the river by paper mills and other industrial operations.

These examples of modern pollution are continuous with much longer histories of environmental degradation wrought by what the anthropologist Anne Spice (2018: 44) calls the "invasive infrastructures" of settler colonialism. Eighteenth- and nineteenth-century settlers in Michigan decimated the sturgeon populations and destroyed the wild rice beds that had long sustained the lifeways of the Anishinaabe people. Early in the twentieth century, hydroelectric dams built to sustain sawmills altered the river's flow and disrupted its ecology. Later, in the 1960s, Michigan experienced its own miniature oil boom; the gushers and seeps and leaks that have long typified petroleum extraction soaked wetlands and tributaries and bled oil into the river. In New Mexico, the war against Native people has likewise long been waged as a war against the land: from the nineteenth-century eradication of peach, corn, and bean plantations as an act of conquest; to the mass slaughter of Navajo sheep, horses, cows, and goats in the 1930s to clear the land for the Hoover Dam; to the oil wells and coal and uranium mines that still occupy the land today. Thus, the histories of the settler occupation of Dinétah and of Michigan are histories of the systemic eradication of life-giving infrastructures and their replacement with eliminationist infrastructures of extraction.

From the perspective of US settler history, these latter infrastructures have long constituted visible signs of progress; they are seen as wonders of technological ingenuity that promise civilizational advancement, wealth, and freedom. Yet this "discursive positioning of infrastructure as a gateway to a modern future" (Spice 2018: 42) obscures the destruction and violence that settler infrastructures impose in their everyday operations as well as at moments of breakdown, which often painfully reveal the uneven distribution of those infrastructures' promised benefits across populations. Promiscuous in their range of meanings and effects, built infrastructures are "dense social, material, aesthetic, and political formations that are critical both to differentiated experiences of life and to expectations of the future" (Appel, Anand, and Gupta 2018: 3). Perhaps nowhere has this been rendered more vividly in the twenty-first century than in the many instances of infrastructures as both instruments of injustice and sites of resistance: from the activists working to secure clean water in Flint and Detroit, to the abolition movement seeking to dismantle brutal police forces and the prison-industrial complex, to the grassroots movement against the Keystone XL pipeline and the Indigenous-led mass protests at Standing Rock against the Dakota Access Pipeline.

The latter protests respond not just to historical injustices, including the ongoing violence inflicted by settler colonialism and extractive capitalism. They respond also to the planetary ecological emergency produced by these long histories: global warming. In the context of that emergency, infrastructures occupy a complex place, at once causes, casualties, and solutions. The infrastructures of empire and capital require and sustain fossil fuel consumption, which is why, for example, new oil and gas pipeline projects have become pivotal sites of protest for grassroots climate activists and Indigenous water protectors. Given the petroleum economy's resistance to democratized labor action, thwarting new pipeline infrastructure construction provides an indirect means of halting extraction.[3] At the same time, already-precarious and deteriorating infrastructures are imperiled by the myriad effects of climate change: rising sea levels threaten coastal power plants; intensified weather events like hurricanes overwhelm water and sewer systems; heat waves cause stress on roads and bridges owing to thermal expansion. The lethal combination of drought-like conditions, settler-colonial land management, and poorly maintained power lines in California, for example, has increased the frequency and intensity of wildfires, destroying homes, lives, and thousands of acres of forest.

Yet infrastructure also promises the solution to these problems.[4] A shift from greenhouse-gas-emitting energy sources to renewables will require massive public and private investments in new infrastructural systems: new modes of mass transportation, wind farms, and solar-powered energy grids. The need for such infrastructure constitutes its own kind of emergency, one that calls for both dismantling and rebuilding. As Michael Truscello (2020: 31) puts it in *Infrastructural Brutalism: Art and the Necropolitics of Infrastructure*, "only a revolutionary examination of existing infrastructure, retrofitting of carbon-intensive infrastructure, decommissioning of untenable infrastructure that requires expert knowledge, and sabotaging of the infrastructure that is too dangerous to keep is capable of transforming conditions currently so brutal that they incur mass extinction." A recently released report from Princeton University (Larson et al. 2020) details the urgent necessity of such a revolutionary transformation. According to the report, the United States will need to act with extraordinary haste to construct the systems that will allow the country to reach net-zero greenhouse gas emissions by 2050—a goal promoted by many, including President Biden. That means an enormous expansion in the nation's electrical grid, an accelerated rate of installation of wind turbines and solar panels, the retrofitting of homes so that they can be powered by energy sources other than oil and gas, and the construction of a new network of currently nonexistent carbon-capture storage facilities, among other efforts.

As sobering as the Princeton study (and the Intergovernmental Panel on Climate Change reports on concentrations of global atmospheric carbon to which it implicitly responds) might appear, the framework of emergency is not without its own dangers. For one thing, its visions of both infrastructural change and the climate emergency might not be expansive enough. For while it imagines new forms of energy production, distribution, and consumption, it fails to imagine new social arrangements; as contemporary movements for ecosocialism, decolonization, and environmental racial justice suggest, infrastructural redevelopment without social and economic reorganization can quite easily perpetuate the immiseration that defines our present. Put another way, heeding the call to swift action on such a broad array of initiatives risks reproducing the ill effects of what the Potawatomi scholar Kyle Powys Whyte (2021) calls "crisis epistemology." Crisis epistemology apprehends present conditions as both new and urgent. The presumption of newness, the unprecedented, allows for a degree of historical forgetting of past emergencies, while the presumption of urgency "suggests that swiftness of action is needed to cope with imminence. There either may be moral sacrifices that have to be made or ethics and justice are not elevated to a level of serious attention" (Whyte 2021). To illustrate the point, Whyte cites historical instances of injustice resulting from infrastructure projects in the twentieth century: dam projects that flooded Indigenous land and displaced Indigenous people, and the exclusion of Indigenous peoples from energy grid systems. With its single-minded focus on emissions reductions and technologies, the Princeton study provides no guidance to policymakers who might take up the report regarding how the pursuit of the study's goals might wind up duplicating, in the name of solving a crisis that is "new" only from a settler perspective, past instances of environmental or infrastructural injustice.[5]

But if it's hard to relinquish "emergency" and "crisis" as responses to climate change, then perhaps thinking differently about infrastructure can help address the epistemological and ethical problem that Whyte identifies. Here, we might turn to Winona LaDuke and Deborah Cowen's (2020: 244) concept of "Wiindigo infrastructure." The infrastructures of settler colonialism disposition the world toward its own destruction, enabling the transformation of the whole planet into an "extractive zone" (Gómez-Barris 2017) for the production of capital and the unevenly distributed immiseration of human and nonhuman life.[6] For LaDuke and Cowen (2020: 244), this aligns settler-colonial infrastructures with the Wiindigo, "the cannibal monster of Anishinaabe legend" that figures the rapacious extractivism of colonial powers. Yet as LaDuke and Cowen also emphasize, the immense scale of Wiindigo infrastructure and its power to determine how things will go does not make it an unstoppable force. Rather, to focus on infrastructure as a way that power is expressed and mediated is to open up a new set of opportunities for subversion, resistance, and counter-world-making: "infrastructure is the spine of the Wiindigo, but it is also the essential architecture of transition to a decolonized future" (246). The redispositioning of infrastructure thus becomes a crucial tool for those seeking to produce different realities in the present and alternative possibilities for the future.

What role might literature and literary studies play in this work of redispositioning? Insofar as it shapes our perception of both space and time, infrastructure might be understood as a structuring form. In this, as Caroline Levine (2015) has explained, it has much in common with literature, itself composed of structuring forms that simultaneously reflect and construct our ideas about the world and how to move through it. In this special issue of *American Literature*, we seek to interrogate the role of literature and literary study in the analysis and reimagining of the infrastructure of emergency. What is the role of literature in helping us think through, around, beyond, and beneath the impasse of infrastructure? How can literature produce imaginaries of infrastructure that might enable us to use it for something other than its current fascistic, colonial, and extractive tendencies? How can literary representations of infrastructure modulate our lived experience of attachment, entanglement, relationality, and collectivity, especially as a way of combating a resurgent fascism that denies any value in public infrastructure beyond the prison, the concentration camp, and the border wall? How can literature help to revivify a commitment to the common good—one that might, for the first time in the United

States, be capable of conceptualizing a public good that is not built on the bodies of those excluded from the public but on mutual aid and an ethics of care? Some version of collectivity is more important than ever, but we have to be able to redefine community as separate from the ideas of national identity and national progress that have led us into the current emergency. What is the relationship between rethinking infrastructure and rethinking collectivities that do not reproduce the eliminationism, homophobia, racism, ableism, gendered violence, and ecocide that have been produced by colonialism?

To think of infrastructure as something that can be retooled for justice is simultaneously a more challenging and a more modest aim than models of revolution built on burning it all down to begin again.[7] If another world is indeed possible, then how can we make it from the infrastructures of the world that we have? In asking these questions, we seek to move infrastructure studies beyond the paradigm of visibility that often frames the problem and its solution in terms of perception. This framework emphasizes what the literary critics Michael Rubenstein, Bruce Robbins, and Sophia Beal (2015: 576) call the "boringness of infrastructure," the fact that it plays an invisible, background role in most of our lives; the fact that in its smooth operation, most of us can take it for granted. The critical approach they call "infrastructuralism" attends to literary texts "that try to make infrastructure, as well as its absence, visible." Elsewhere, Robbins (2007: 29), for example, thus takes up the appearance of public utilities in fiction by Jonathan Franzen, Milan Kundera, and others in order to bring them into focus as "objects of political struggle." And Levine (2015: 588) considers material infrastructures in relation to other deep structural formations like racial hierarchies, in order to "unsettle the privileged obliviousness" that obscures the work that infrastructure performs. This focus upon the taken-for-granted quality of infrastructure is echoed in the work of Kate Marshall (2013: 82), who has argued that the "becoming-visible of infrastructure" in the US modernist novel helps "reveal the novel's communicative architectonics" (84).

Yet scholars working in other fields, like anthropology and cultural geography, have already begun to reveal the limitations of the visibility/invisibility paradigm. As Thomas S. Davis (2021), following thinkers in these fields, has argued, "we need not only make infrastructure visible, but to enact ways of seeing it better." The history and literature of the United States provide an especially rich archive of materials for a critical project that seeks to trace the genealogies and theorize the imagined futures enabled (or foreclosed) by infrastructure. Since settler arrival, those infrastructures—from the transcontinental railroad to the interstate system, from coal mines to solar farms—have served as appurtenances of empire and capital *and* as signs of progress; they have been instruments of violence, destruction, oppression, and ecological degradation *and* tools of wealth and growth. In ways as yet largely unexamined, US poetry and narrative fiction have documented and contested the social, cultural, and ideological work performed by these infrastructures, often "interrogat[ing]" as Jessica Hurley (2020: 3) shows in her study of US nuclear infrastructure, "the racist, sexist, colonial, and homophobic logics that structured" them.

The contributions to this special issue initiate just such a series of investigations. In "War on Dirt: Aesthetics, Empire, and Infrastructure in the Low Nineteenth Century," Andrew Kopec frames the settler-colonial project as one organized by the imperative to transform dirt into infrastructure. Tracing the literary history of water infrastructures such as the Panama and Erie Canals, Kopec argues that the two primary aesthetics of infrastructure—the sublime drive toward modernity described by Appel, Anand, and Gupta and the everyday boring hiddenness described by Rubenstein, Robbins, and Beal—in fact operate dialectically to facilitate settler colonialism's seizure and transformation of the landscape: large national infrastructure projects are seen first as too sublime to be stopped and then as too mundane for their ongoing violence to be registered. Cutting through an opposition that has long defined critical infrastructure studies, Kopec reveals instead how the spatiotemporal politics of infrastructure change over time according to the needs of the settler-colonial state. And yet the capacity for shifts in how infrastructure signifies also holds open the possibility that new ways of inhabiting infrastructure—and especially the ruins of infrastructure that mark our contemporary emergencies— might yet develop, allowing us to move beyond the "infrastructural dialectic" and create "a platform from which to reject the appetitive, imperial wild surmises that run through the infrastructures of coloniality from the past to the present."

If affect organizes how we perceive and respond to infrastructure, as Kopec suggests, then literature becomes a crucial mechanism for modulating larger structures of feeling around infrastructure and the public good. Jamin Creed Rowan's "The Hard-Boiled Anthropocene and the Infrastructure of Extractivism" proposes hard-boiled fiction and the noir tradition as unexpected sites where such a modulation occurs, showing how the obliviousness and cruelty that define the noir villain map on to the victory of extractivist infrastructures from the dam to the oil field. By redefining the infrastructures of extraction as "a type of weapon through which individuals commit crimes" and "a mechanism for concealing criminality," hard-boiled fiction, composing a structure of feeling in which revelation leads to shock and outrage, subverts the "emotional and ethical evasion" that infrastructure facilitates. More recent hard-boiled climate fiction such as Paolo Bacigalupi's *The Water Knife* (2015) builds on this tradition to show how both the present and the future have been immiserated as sources of value extraction via the infrastructures of the past, a grim vision but one that, nonetheless, allows for the emergence of new collectivities in the ruins of the past's collapsing infrastructures. Building on AbdouMaliq Simone's theory of "infrastructures of relationality," Rowan proposes that the collective maintenance of infrastructures in the service of communal survival might recompose an idea of the common good that would be oriented toward life rather than extraction.

The noxious notions of individual sovereignty that define the noir villain are also central to Suzanne F. Boswell's "'Jack In, Young Pioneer': Frontier Politics, Ecological Entrapment, and the Architecture of Cyberspace." Boswell demonstrates that the arrival of the internet in the United States was deeply entangled with ecological emergency, as the supposed immateriality of cyberspace was seen as offering an escape from a world running out of space and resources. Immediately mapped on to the older frontier chronotope, cyberspace became a disembodied place where whiteness, sovereignty, and individualism reigned and to which "the white user belonged as an indigenous inhabitant." These early colonial and ecological frameworks for understanding virtual space would go on to shape the technological and digital infrastructures of the internet through the 1990s, "influencing the modern user's experience of the internet as a private space under their sovereign control." If for Kopec infrastructure materializes empire, then Boswell demonstrates how ideas of empire and conquest have shaped our ostensibly immaterial digital infrastructures, foreclosing the possibility of an internet defined by collectivity in favor of one that demands "individual and private access."

And yet, as Boswell demonstrates in her reading of William Gibson's Sprawl trilogy (1984–1988), the ecosystem of cyberspace never succeeds in manifesting the new frontier. Rather, it places users in an inescapable infrastructural connection to everything the frontier would prefer to avoid: the collective, the nonsovereign, the political, and particularly the racialized other. Here we see infrastructure operating as something like the collective unconscious of frontier individualism, materializing the political and social connections and the intimacy between bodies that the image of the cowboy is meant to preclude. Boswell's work thus offers a way to cut through one of the key impasses of the present emergency: the ceaseless demand by those in power that we develop individual solutions to planetary emergencies like climate change. Attention to the infrastructures of climate crisis is not only a more accurate way of perceiving cause and effect, but it is also a way of accessing the reality of collective agency in the Anthropocene without falling into the flattened universalism of what Dipesh Chakrabarty (2009: 213) has called "species think-ing." Infrastructure is collective and puts us into relation whether we like it or not; it also shows us *how* we are related, to what degree, and within what flows of power. Infrastructure mediates the univer-sal and the individual, offering a systemic analysis that maps con-nections, causes, and consequences across different scales of space and time without suggesting that "we" are all equally culpable for what unfolds.

The first three essays in this collection show how settler capitalism has shaped the infrastructural transformation of the continent, from the canal to the data center. The final three essays focus on the inter-sections of infrastructure with the ongoing emergencies of racial oppression: infrastructure as racial violence. In "Geomemory and Genre Friction: Infrastructural Violence and Plantation Afterlives in Contemporary African American Novels," Rebecca Evans considers the spatial dimensions of what Toni Morrison calls rememory: the simultaneously gothic and mimetic hauntings that are produced when the violent racial histories of place continue to create racial violence in the present through the continuities of infrastructure. Evans pro-poses the term *geomemory* to define an emergent formal technique in contemporary African American fiction where the racial emergencies of the present are revealed to be the result of still-active infrastructures of white violence; what might be taken to be a purely gothic form is theorized by Evans instead as one produced by the friction between gothicism and realism, where the haunting of the present by the past is less a suggestive trope than a literal reality. As Evans writes, "geo-memory doesn't just show that history haunts us; it shows precisely how that haunting plays out in the human use and misuse of land and the human organization of space and infrastructure." Tracing the afterlives of the plantation to the petrochemical plants that debilitate bodies in Louisiana's Cancer Alley, the prison farms that Jesmyn Ward represents in *Sing, Unburied, Sing* (2017), and the postbellum medical experiments evoked by Colson Whitehead in *The Underground Railroad* (2016), Evans shows how the temporal collapses of Black gothicism and geomemory serve to puncture the aggressive forward drive of infrastructure-as-national-modernity-and-concretized-sign-of-progress, revealing and contesting instead the temporal drag by which infrastructure situates racialized bodies within the ongoing catastrophe of white supremacy.[8]

If the plantation lives on in the social and spatial continuities of its infrastructures, how, then, might we imagine modes of survival and flourishing for Black lives in worlds that are infrastructured toward their containment and destruction? In "The Subsident Gulf: Refiguring Climate Change in Jesmyn Ward's Bois Sauvage," Kelly McKisson reorients our vision away from the infrastructures that occupy the ground and toward the ground itself. McKisson shows how large national infrastructure projects in the Gulf have led to the loss of land as an ecological infrastructure, as leveeing, river channeling, and pipelines, as well as the rising waters of climate change, produce accelerated subsidence that disappears land at a rate of an acre an hour in the Mississippi Delta. For McKisson, subsidence becomes visible in Ward's work as a figuration that binds environmental crisis to environmental and racial injustice, with "sinking" figuring both the disappearing land and the literally and metaphorically drowning bodies that Ward invokes in her representations of Black Gulf life during and after Katrina. And yet, the search for dry ground that defines nationalist responses to climate change also loses its power. Rather, the question in Ward's novels is how to live on shifting, sinking, vanishing land; how to orient toward water; how to salvage a life at the waterline; how to inhabit and imagine space differently as a form of resistance to the subsident force of infrastructural emergency. Security and stability may not be possible, here, but they also may not be desirable in a world structured by infrastructural violence. Infrastructure promises continuity across time, but that continuity can also mean the continuation of racial and ecological violence. In improvisational acts of salvage that seek to prolong subsistence in a subsiding world, McKisson sees how "the instability of the land can be countered by the peoples' resistant and imaginative postures of dissension." A subsiding world, a world of ongoing emergency, requires an infrastructure built anew each day, a place that is composed and recomposed by dissident acts in constantly renewing relation.

We close this issue with an essay that transitions us away from more traditional infrastructure projects to one that is less immediately visible as such: the conglomeration of buildings, laws, institutions, and capital markets that is the US healthcare system. Michelle N. Huang's "Racial Disintegration: Biomedical Futurity at the Environmental Limit" argues that the "underlying conditions" of health "extend beyond the individual racialized body and should refer also to the structural and material conditions of damage to which it is subject." Huang challenges the deracination of healthcare—which promises a postracial utopia of individually tailored genomic medicine while ignoring all of the ways in which racialized environmental emergencies render people unevenly vulnerable to death (Gilmore 2007)—by drawing on contemporary Asian American dystopian literature, which "provides a crucial case study through which to analyze futurities where healthcare infrastructures intensify racial inequality under terms that do not include race at all." Understanding health as a consequence of space and infrastructure contests the privatization of medicine, with its emphasis on individual wellness and personalized solutions, and suggests that race, solidified in infrastructure and infrastructural violence, will manifest as damage to bodies that we perceive as illness rather than recognizing it as the product of racial violence. For Huang, the technique of "studious deracination" used in recent novels by Chang-rae Lee and Rachel Heng entrains us to see how racism works "outside-in, not inside-out," with race adhering in and emerging from the infrastructures that white supremacy produces.

As we write in the early days of 2021, the vexed relations between infrastructures and emergencies of various kinds are perhaps as vivid as ever—electoral systems and democracy, the healthcare system and pandemic, the state carceral apparatus and racial violence, extractive infrastructures and climate change. Meanwhile, the specters of Church Rock and Marshall loom as an outgoing presidential administration approves new uranium mining on public lands in the West and Enbridge embarks upon massive new pipeline infrastructure projects in the Midwest—each in the name, perversely, of progress and security. This continuation of the very practices that have produced our present crises prompts us to think not just in terms of the infrastructure *of* emergency but also in terms of infrastructure *as* emergency. The essays gathered here and in our companion special issue of *Resilience: A Journal of the Environmental Humanities* document several iterations of that convergence, helping us to grapple with infrastructure's position at the heart of struggles over world-making from the eighteenth century to the present. In these essays, infrastructure comes alive as a site of attachment, a form of relation, and a shifting improvisation toward a future that the present's infrastructures of/as emergency do not entirely foreclose. But if infrastructure is always, as Anne Spice (2018: 47) observes, "assembled in the service of worlds to come," those worlds need not be the ones imagined by racial capitalism with its insatiable appetite for progress and growth at the expense of justice and reciprocity. We live in a time in which the things that bind us are shifting and many of us experience this, rightly, as an emergency. But LaDuke and Cowen remind us that, on the ground, those who have been harmed by or excluded from the benefits and advantages of "Wiindigo infrastructure" continue, as they always have done, to create communities of care, to imagine and build infrastructures otherwise.

Jessica Hurley is assistant professor of English at George Mason University, where she teaches speculative fiction, critical theory, and the environmental humanities, and the author of *Infrastructures of Apocalypse: American Literature and the Nuclear Complex* (2020). She is currently working on her second book project, "Nuclear Decolonizations."

Jeffrey Insko is professor of English at Oakland University in Michigan, where he teaches courses in nineteenth-century American literature and culture and the environmental humanities. He is the author of *History, Abolition, and the Ever-Present Now in Antebellum American Writing* (2019). He is currently completing a book titled "Untimely Infrastructure: The 2010 Marshall, Michigan Oil Spill in the Human Epoch."

Notes

1   For an overview of the damage and a critique of the cleanup so far, see Fettus and McKinzie 2012.

2   For a full account of the Enbridge spill, see "The Dilbit Disaster" (2012), a series of articles published by reporters Elizabeth McGowan, Lisa Song, and David Hasemyer in *Inside Climate News*. The series won the Pulitzer Prize in 2013. https://insideclimatenews.org/news/26062012 /dilbit-diluted-bitumen-enbridge-kalamazoo-river-marshall-michigan-oil -spill-6b-pipeline-epa/.

3   We have in mind Timothy Mitchell's well-known account in *Carbon Democracy: Political Power in the Age of Oil* (2011).

4   For an account and critique of infrastructure's promises, see Appel, Anand, and Gupta 2018.

5   For additional Indigenous critiques of crisis and apocalypticism, see Simpson 2017 and Mitchell and Chaudhury 2020.

6   Keller Easterling (2014) uses the term "disposition" to name infrastructure's capacity to determine how things will go; infrastructure can be dispositioned to produce certain outcomes, and perhaps redispositioned to produce others.

7   As Reuben Martens and Pieter Vermeulen (2021) note in their essay in
     the issue of *Resilience* that makes up the other half of this special issue,
     imagining infrastructure as a form of continuity that can get us to the
     future that we want is an important alternative to the impasse of infra-
     structure that can see only its immovable rigidity or its spectacular
     destruction.
8   We take the term *temporal drag*, of course, from Elizabeth Freeman
     (2010: 62), who uses it to index "the pull of the past on the present."
     While for Freeman temporal drag is (or can be) a reparative practice in
     queer life, we see (in our reading of Evans's work) infrastructure as a
     materialization of temporal drag that binds Black and other dispossessed
     subjects to the ongoing conditions of violent pasts.

References

Appel, Hannah, Nikhil Anand, and Akhil Gupta. 2018. "Introduction: Tempo-
    rality, Politics, and the Promise of Infrastructure." In *The Promise of Infra-
    structure*, edited by Nikhil Anand, Akhil Gupta, and Hannah Appel, 1–38.
    Durham, NC: Duke Univ. Press.
Brugge, Doug, Jamie L. deLemos, and Cat Bui. 2007. "The Sequoyah Corpora-
    tion Fuels Release and the Church Rock Spill: Unpublicized Nuclear
    Releases in American Indian Communities." *American Journal of Public
    Health* 97, no. 9: 1595–1600.
Chakrabarty, Dipesh. 2009. "The Climate of History: Four Theses." *Critical
    Inquiry* 35, no. 2: 197–222.
Davis, Thomas S. 2021. "'Far from the Gulf Coast, But Near it Too': Art,
    Attachment, and Deepwater Horizon." *Resilience: A Journal of the Envi-
    ronmental Humanities* 8, no. 3.
Dempsey, Dave. 2001. *Ruin and Recovery: Michigan's Rise as a Conservation
    Leader*. Ann Arbor: Univ. of Michigan Press.
Easterling, Keller. 2014. *Extrastatecraft: The Power of Infrastructure Space*. Lon-
    don: Verso.
Environmental Protection Agency (EPA). n.d. "United Nuclear Corp. Church
    Rock, NM: Cleanup Activities." https://cumulis.epa.gov/supercpad/Site
    Profiles/index.cfm?fuseaction=second.cleanup&id=0600819 (accessed
    January 8, 2021).
Fettus, Geoffrey H., and Matthew G. McKinzie. 2012. "Nuclear Fuel's Dirty
    Beginnings: Environmental Damage and Public Health Risks from Ura-
    nium Mining in the American West." *Natural Resources Defense Council*,
    March. https://www.nrdc.org/sites/default/files/uranium-mining-report
    .pdf.
Freeman, Elizabeth. 2010. *Time Binds: Queer Temporalities, Queer Histories*.
    Durham, NC: Duke Univ. Press.
Gilmore, Ruth Wilson. 2007. *Golden Gulag: Prisons, Surplus, Crisis, and Oppo-
    sition in Globalizing California*. Berkeley: Univ. of California Press.

Gómez-Barris, Macarena. 2017. *The Extractive Zone: Social Ecologies and Decolonial Perspectives*. Durham, NC: Duke Univ. Press.

Hurley, Jessica. 2020. *Infrastructures of Apocalypse: American Literature and the Nuclear Complex*. Minneapolis: Univ. of Minnesota Press.

LaDuke, Winona, and Deborah Cowen. 2020. "Beyond Wiindigo Infrastructure." *South Atlantic Quarterly* 119, no. 2: 243–68.

Larson, Eric et al. 2020. *Net-Zero America: Potential Pathways, Infrastructure, and Impacts* (interim report). Princeton University, December 15. https://environmenthalfcentury.princeton.edu/sites/g/files/toruqf331/files/2020-12/Princeton_NZA_Interim_Report_15_Dec_2020_FINAL.pdf.

Levine, Caroline. 2015. *Forms: Whole, Rhythm, Hierarchy, Network*. Princeton, NJ: Princeton Univ. Press.

Marshall, Kate. 2013. *Corridor: Media Architectures in American Fiction*. Minneapolis: Univ. of Minnesota Press.

Martens, Reuben, and Pieter Vermeulen. 2021. "Infrastructural Prolepsis: Contemporary American Literature and the Future Anterior." *Resilience: A Journal of the Environmental Humanities* 8, no. 3.

Mitchell, Audra, and Aadita Chaudhury. 2020. "Worlding Beyond 'the' 'End' of 'the World': White Apocalyptic Visions and BIPOC Futurisms." *International Relations* 34, no. 3: 309–32.

Mitchell, Timothy. 2011. *Carbon Democracy: Political Power in the Age of Oil*. London: Verso.

National Transportation Safety Board. 2012. *Enbridge Incorporated Hazardous Liquid Pipeline Rupture and Release, Marshall, Michigan, July 25, 2010*. Pipeline Accident Report NTSB/PAR-12/01. Washington, DC. https://www.ntsb.gov/investigations/AccidentReports/Reports/PAR1201.pdf.

Robbins, Bruce. 2007. "The Smell of Infrastructure: Notes toward an Archive." *Boundary 2* 34, no. 1: 25–34.

Rubenstein, Michael, Bruce Robbins, and Sophia Beal. 2015. "Infrastructuralism: An Introduction." *Modern Fiction Studies* 61, no. 4: 575–86.

Simpson, Leanne Betasamosake. 2017. *As We Have Always Done: Indigenous Freedom through Radical Resistance*. Minneapolis: Univ. of Minnesota Press.

Spice, Anne. 2018. "Fighting Invasive Infrastructures: Indigenous Relations against Pipelines." *Environment and Society* 9, no. 1: 40–56.

Truscello, Michael. 2020. *Infrastructural Brutalism: Art and the Necropolitics of Infrastructure*. Cambridge, MA: MIT Press.

Voyles, Traci Brynne. 2015. *Wastelanding: Legacies of Uranium Mining in Navajo Country*. Minneapolis: Univ. of Minnesota Press.

Whyte, Kyle Powys. 2021. "Against Crisis Epistemology." *Handbook of Critical Indigenous Studies*, edited by Brendan Hokowhitu, Aileen Moreton-Robinson, Linda Tuhiwai-Smith, Steve Larkin, and Chris Andersen, 52–64. London: Routledge.

Andrew Kopec

War on Dirt:
Aesthetics, Empire, and Infrastructure
in the Low Nineteenth Century

Abstract This essay considers the politico-aesthetics of infrastructure by focusing on poems that anticipate, justify, and critique internal improvements, from Joel Barlow's early Republican vision of the Erie and Panama Canals to texts that document the ruin caused by the works Barlow imagined as glorious. Historical scholarship has long assessed the mania for cutting roads and canals into the landscape. But engaging an emerging infrastructuralism—and turning to imaginative texts that exist underneath the ground typically trod by US literary studies, from Philip Freneau's celebratory ode to the Erie Canal to Harriet Beecher Stowe's and Nathaniel Hawthorne's ironic canal travel sketches to Margarita Engle's recent historical verse-novel tallying the devastations of the Panama Canal—this essay identifies an infrastructural dialectic in which writers view infrastructure, initially, as awesome so as to justify its ecological and social violence and, subsequently, as banal so as to render it invisible within the settler state. Oscillating between awe and irritation, the sublime and the stuplime, then, these texts both expose the rhythm of infrastructure's long—that is, low—relation to the structure of coloniality and, in Engle's case, model how to disrupt it so as to imagine a more just life "after" infrastructure.
Keywords infrastructure, sublime, Erie Canal, Philip Freneau, Nathaniel Hawthorne

The Infrastructural Dialectic: The Sublime and the Stuplime

In his nine-book epic, *The Vision of Columbus* (1787), the early Republican poet Joel Barlow vindicates Christopher Columbus as "the Sage . . . / Who taught mankind where future empires lay" (25). Linking Columbus's voyages to an "imperial pan-American imaginary" (Wertheimer 1999: 74), Barlow (1787: 26) depicts how an angel justifies Columbus's "painful years and persevering toil" as expanding a hemispheric empire "o'er the pathless main." And central to this expansion is infrastructure—alterations to the landscape that would make it possible to "trace new regions o'er the" terra firma of the Americas' "bounteous shore" (26). For with "canals, long-winding," a

American Literature, Volume 93, Number 3, September 2021
DOI 10.1215/00029831-9361223   © 2021 by Duke University Press

modern civilization could even "join" the "Hudson [River] to broad Ohio's wave" (246).[1] Ignoring water infrastructure as an instance of "planned violence"[2]—ignoring the injustices to the environment and to Indigenous peoples that its realization would by design entail—the poet insists that this wedding of the waters would be doubly "peaceful" (245). Such a canal would transpire quickly, without "pain," and signal the triumph of "commerce . . . o'er the rage of war" (244).

Barlow's poem captures how infrastructure manifests the relation between empire and ecological disruption. More specifically, it reveals the 363-mile Erie Canal that would connect Lake Erie and the Hudson River as an emblem of coloniality, a term developed by writers such as Sylvia Wynter and Walter Mignolo to analyze "modernity, capitalism, and coloniality [as] aspects of the same package of control" that limits the sense of what lives are "possible" (Mignolo 2007: 163).[3] As part and parcel of settlement's "ceaseless expansion" (Wolfe 2006: 395), the artificial waterways of canals played an important role in this "package of control": they were "schemes" of "enterprising individuals" to improve the land (Livingood 1941: 133) for commerce. Distinguished from its early Republican predecessors by its scale and public financing, the Erie Canal seemed to realize Barlow's expectations for both the symbolic and practical dimensions of this imperial infrastructure. Started in 1817 and completed in 1825, the canal hastened settlement of the American West, sprouted towns and commerce along its banks, and helped transform New York City into a financial capital by providing the United States' first two-way transportation avenue for commerce through its port. The canal's success as an emblem of progress, however, was hardly guaranteed.[4] For even though proposals for the Erie Canal circulated during the canal mania of the late 1700s in the Anglo-American world, construction did not begin until the 1810s. In fact, debates over funding for "internal improvements"—the early Republican term for "an integrated network of roads and canals" (Larson 1987: 363)—were mired in a "long dispute" over "infrastructure development" at both federal and state levels (Hostetler 2011: 54). Yet after New York established the Canal Fund in 1817 (Sheriff 1996: 21), the culture celebrated in 1825 the completion of the Erie Canal. Whereas in the late 1810s, political opponents of DeWitt Clinton, the New York politician who orchestrated the project, had ironically termed the canal "Clinton's Ditch," by the late 1820s, the culture hailed the canal as evidence of "The March of Progress in America" (Woollard 1937: viii).

Previous engagements with water infrastructure have often turned to the Panama Canal, which when completed in 1914 provided a forty-eight-mile-long interoceanic shipping path that realized another dream of colonialists. The imperial dimensions of Theodore Roosevelt and his steam shovel in Colón, Panama (the Spanish name for Christopher Columbus given to the northern terminus of the canal), have been obvious to Americanists since Amy Kaplan and Donald E. Pease's *Cultures of United States Imperialism* (1993), a paradigm-shifting volume that uses the Hercules poster promoting the 1915 Panama–Pacific International Exposition, San Francisco, as its cover (see fig. 1). This image not only visually links empire to white masculinity (Markwyn 2016) but also relays an understanding of US imperialism as a post-1893 projection of power beyond the North American frontier. Yet, as the name of the city Colón beckons us to do, this essay finds that the image should be understood within a longer historical context: in Kyle Powys Whyte's (2017a: 156) terms, as "another intensified episode of colonialism that opens up Indigenous territories for capitalism and industrialization."[5] If the Erie Canal looks backward to the conquest of the Americas, it is also a site of ongoing contest[6] over transforming New World dirt into infrastructure that the Panama Canal, in its facilitation of global trade, extends into the twentieth and twenty-first centuries.

In what follows, I bring forth the politico-aesthetic parameters defining the struggle to transform the "settlerscape."[7] In so doing, my essay makes legible not a long but a low nineteenth century—a body of texts that, in grappling with the social, ecological, and, finally, aesthetic problems of water infrastructure, complement Patrick Wolfe's (2006: 388) insight that "invasion is a structure not an event." According to a twentieth-century report on the Erie Canal by the US Geological Survey, "the settlers sought not beauty from the canal but cheap transport for their produce and their supplies" (Langbein 1976: 59). And yet whereas in *Democracy in America* (1835), Alexis de Tocqueville observed that for antebellum Americans "cutting a canal" was a "momentous political question" (Tocqueville 2009: 302), the texts I turn to across the long nineteenth century found "cutting a canal" to raise momentous aesthetic questions, as well. In order to justify infrastructure's "rage of war" on ecology and Indigenous societies, early Republican writers first excavated the poetry beneath—"infra"—the landscape, a logical sequence canonized by the English Romantic poet John Keats's sonnet, "On First Looking into Chapman's Homer" (1817). With a mistaken reference to Hernán Cortés, rather than Vasco Núñez de Balboa, the sonnet likens reading a translated epic to

Figure 1 Perham Wilhelm Nahl's lithograph *The Thirteenth Labor of Hercules.* Created for the Panama–Pacific International Exposition, San Francisco, 1915, this poster was selected as the exposition's official image. 100 Years: Panama–Pacific International Exposition 1915–2015 (website)

transforming the Panamanian isthmus into an imperial project: the reader is "like stout Cortez when with eagle eyes / He star'd at the Pacific—and all his men / Look'd at each other with a wild surmise— / Silent, upon a peak in Darien" (Keats 1978: 64). Dreaming of what would become the Panama Canal, here Keats voices the "wild surmise" of infrastructure in general: the appetitive, anticipatory longing for what a recent infrastructuralist calls the settlerscape's "not-yet-achieved future" (Gupta 2018: 63).

Romantic historians of the United States have followed Keats's example in sensing this awed, anticipatory sublime as integral to empire's spatio-temporality. The prolific writer Archer B. Hulbert's *The Paths of Inland Commerce* (1920) plucks a moment from George Washington's correspondence from the mid-1780s in which, standing in the upstate New York forest, the American general recalls:

> I could not help taking a more extensive view of the vast inland navigation of these United States and could not but be struck by the immense extent and importance of it . . . Would to God we may have

wisdom enough to improve them. I shall not rest contented till I have explored the Western country, and traversed those lines, or a great part of them, which have given bounds to a new empire. (Washington quoted in Hulbert 1920: 6)

Hulbert finds it simply an "interesting fact" that Washington "should have his first glimpse of this vision from the strategic valley of the Mohawk" (7). The connection between place and desire runs deeper, though: it points to the wild imperial surmise of infrastructure ("could not help," "could not but be struck"), in which the settler imaginatively "improve[s]" "immense" land into a "new empire." Whether in Barlow's verse or Washington's letters, this apprehension of an imperial emblem thus preceded the completion of the Erie Canal, the poet's vision anticipating the thrust of the ditch-digger's shovel.

This discursive connection among aesthetics, infrastructure, and imperialism would extend forward into United States history. Andrew Jackson's infamous Second Annual Message to Congress in 1830, for example, linked his catastrophic removal policies to internal improvements. There he asks rhetorically:

What good man would prefer a country covered with forests and ranged by a few thousand savages to our extensive Republic, studded with cities, towns, and prosperous farms embellished with all the improvements which art can devise or industry execute, occupied by more than 12,000,000 happy people, and filled with all the blessings of liberty, civilization and religion? (Jackson 1837: 114)

Here Jackson puts on offer what Pease (1984) terms "sublime politics," which yoke Manifest Destiny to an ideological process of unearthing a script of "improvements" immanent to the settlerscape. Collapsing distinctions between art and nature, republic and empire, war and blessings so as to divide "savages" and modern "people," here Jackson articulates how "settler colonialism destroys to replace" (Wolfe 2006: 388), and he frames this violence as an ethic in which "good" Americans accept the destruction of ecologies and Indigenous societies as the costs of empire.

Like Washington, Jackson's discourse clearly wants to ennoble the settler mission, but the fact that he also couches this view of "improvements" in banal demographic terms ("12,000,000 happy people") points to a paradox manifested in the poems, addressed in the next part of the essay, by Philip Freneau and Samuel Woodworth that celebrate the shovel or the "American axe"—James Fenimore Cooper's term in

The Crater (1847) for "that glorious implement of civilization" (Cooper 1863: 232).[8] This paradox has occupied recent critical infrastructuralism. In their essay, "Infrastructuralism: An Introduction" (2015), Michael Rubenstein, Bruce Robbins, and Sophia Beal address the tension between experiencing infrastructure on the one hand as sublime and on the other one as banal. In terms of the former, they write: "Because of this vastness, infrastructure tends to have the same stupefying effect as the Kantian sublime" (Rubenstein, Robbins, and Beal 2015: 576). And yet, they continue, "routinely to contemplate" the complexity of (say) a "giant coal-burning electricity plant" "would be to risk derailing your entire sense of purpose." From this perspective, the "inherent boringness of infrastructure" makes "it possible to get on with the everyday activities of everyday life."

These two separate, seemingly paradoxical responses to modern infrastructure—awe and boredom—actually constitute a dialectic in which infrastructure figures, initially, as a product of shocking violence and, subsequently, of everyday life. This rhythm invisibilizes infrastructure's violence: if one perceives infrastructure as boring so as "to get on with" life, one is unlikely to consider it as part of a campaign that facilitated expropriation and settlement from the fifteenth to the twenty-first centuries.[9] Into the nineteenth century this sense of infrastructure's dialectic, like the water in the Erie Canal, would seep through its container. For Harriet Beecher Stowe and Nathaniel Hawthorne, infrastructure provoked a philosophic ambivalence, if not outright ridicule. The third part of my essay considers two texts that pause over infrastructure to imagine the politico-aesthetic costs of the Jacksonian mandate to improve the landscape: Stowe's "The Canal Boat" (1841) and Hawthorne's "The Canal-Boat" (1835), which reveal the cracks in the era's sublime politics. In their ambivalence toward infrastructure, they show the possibility for what we might call, following Sianne Ngai's theorization of the "stuplime" in *Ugly Feelings* (2005), a stuplime politics instead.

As much as Stowe and Hawthorne might contest the Erie Canal's glory, their full-blown critique of infrastructure's politico-aesthetics never materializes, a failure that Margarita Engle's historical verse-novel for young people, *Silver People* (2014), aims to redeem. In a brief coda devoted to this novel, I follow Barlow's (1787: 143n) wild surmise from Lake Erie to the "Isthmus of Darien," so as to connect the Panama Canal to a longer history of coloniality's ecological and social crises that feed the state of emergency in the present. Like recent decolonial work in the environmental humanities by Heather

Davis and Zoe Todd (2017: 765), Engle's novel shows how modern infrastructure is always "already entwined with colonialism." The novel seems of a piece with Davis and Todd's efforts to "train our imaginations to the ways in which environmental destruction has gone hand in hand with colonialism" (769). At the same time, Engle's novel goes further than implying the Panama Canal as an emblem of the Anthropocene. It models how to break the very infrastructural dialectic that leaves Washington awed in the upstate forest and Hawthorne immobilized in the muck of the Erie Canal: historicizing the Panama Canal's actual and epistemic violence, the novel capitulates to neither the sublime nor the banal.

With this refusal, *Silver People* deconstructs the techno-racial triumphalism of Roosevelt and imagines infrastructure as a site of provision for cross-species, cross-racial assemblages that can emerge from its devastations. It imagines life "after" infrastructure. Written a few years prior to Donald Trump's (2018) infrastructure initiative, which promises to "make American infrastructure . . . the envy of the world," Engle's text blunts the cycle of the infrastructural sublime that Trump wishes to renew. Atop an island that used to be a mountain, her young characters desire not domination over the isthmus's ecology but an understanding of it. In doing so, they aim "productively [to] disrupt and undo [the] universalizing and violent logics" (Davis and Todd 2017: 765) celebrated in early Republican poetry and incompletely unraveled in antebellum writing. If Engle's characters read beneath the settlerscape, they do so to adapt its altered politico-aesthetics for a more equitable infrastructure now.

Odes on the Grand Canal

If Barlow imagined the Erie Canal ex nihilo, other bards sang of the campaign in medias res. In doing so, they revealed how the aesthetic discourse of internal improvements linked an altered landscape to an altered sense of the sublime. As the epigraph for his ode, "On the Great Western Canal of the State of New York" (1822), Freneau, an anti-Federalist poet-editor, selected a passage from the ancient poet Horace. In the translation he provides, the verse poses the rhetorical question, "Which was best—to travel through tedious, / dreary forests, or to sail on these recent waters?" (Freneau 1993: 3). Here, Freneau points to a paradox of "recent waters" excluded from the Kantian sublime. As Ngai reminds us, Immanuel Kant had distinguished between the mathematical and the dynamical sublime: he referred, in the first instance, to an affective quality of an object that stems from its magnitude or scale and, with the second one, to an object's quality of terrible, potentially harmful force. Crucially, according to Ngai (2005: 265), Kant limits the experience of the sublime to objects of "rude nature," a limitation that "explicitly bars [the sublime] from being applied to products of art [in Kant's words from the *Critique of Judgement* (1790)] 'where human purpose determines the form and size.'" From landscape painters to poets, artists turning to US infrastructure defied this prohibition by seeing artificial objects such as canals, bridges, and locks as sublime. And in the process of celebrating "recent waters," they turned the Erie Canal into one of the first icons of what the historian David E. Nye (1994) terms the "American technological sublime."

With this alteration to aesthetic theory, Americans enacted a "double action of the imagination by which the land was appropriated as a natural symbol of the nation while, at the same time, it was transformed into a man-made landscape" (Nye 1994: 37). It aimed to upturn the landscape, accommodating the mandate of Manifest Destiny in order to realize the land's fullest, immanent potential according to its own "script" (39). Freneau's "On the Great Western Canal" evinces these sublime politics by poeticizing the Erie Canal as fulfilling Nature's will.[10] Such boosterism for internal improvements might have surprised his anti-Federalist associates. A vehement Democrat-Republican who edited the *National Gazette* (1791–1793), the partisan paper initiated by Thomas Jefferson and known for its vituperative attacks on Alexander Hamilton, Freneau (1963a, 2:160) had previously spoken out against the "brazen age" that the Federalists' wide-ranging national program represented. In his earlier poem, "The Projectors" (1782), the speaker, yearning for "Rome, of old," imagines a prior era in which man "His life in Nature's affluence spent" until "Jobbers" ushered in a new "ruling passion" propelled by ambition and "scheme[s]" (Freneau 1963a, 2:160–61). Before the schemers' imagined subversion of the Republic, "People were then at small expence, / They dug no ditch, made no fence" (2:160). Contrary to the "Base grasping souls" who improved the land in pursuit of schemes, "on a small and scanty spot, / With much ado his living got, / Inured to labour from his birth / Each Roman soldier tilled the earth" (2:161). The projects of internal improvements—digging ditches and building fences, rather than simply "till[ing]" one's affluent but "scanty spot"—were thus seen dangerously to extend the Federalist ethos to the values, tempo, and spatial organization of everyday life.[11]

This anti-Federalist commitment to agrarianism seemed predisposed against the sublimity of internal improvements, but the romance of infrastructure would bridge partisan boundaries,[12] as it led ideological opponents like Freneau and Hamilton to find beneath the landscape a romance written by nature itself. Whereas Jefferson had demurred when approached in 1809 by the New York state delegation for federal funding for the Erie Canal,[13] in his *Report on Manufactures* (1791), Hamilton imaginatively linked the nation's commercial development to "meliorations" of its infrastructure "facilities" (Hamilton 1827: 58). For Hamilton, in a sentiment that would carry the day in New York, these tentative "symptoms of attention to the improvement of inland navigation . . . must fill with pleasure every breast warmed with a true zeal for the prosperity of the country" (59).

Infrastructure thus emerged as a platform on which to build—and thereby delimit—a national future in which "pleasure" was attached to the "improvement" of the landscape. As this romance prevailed in the early United States, what the poet proscribed in the 1780s, he celebrated in the 1820s. "Doing what Nature commanded" (Pease 1984: 278n9), he helped build consensus over the "general utility of artificial roads and canals" (Gallatin 1968: 5). Doing so shows the extent to which a spatially, temporally, and emotionally restricted political imaginary could coalesce around infrastructure. For the editors of *The Promise of Infrastructure*, "infrastructures . . . served to permit states to separate politics from nature" (Appel, Anand, and Gupta 2018: 4). Yet a poem like Freneau's actually wants to depoliticize infrastructure by *collapsing* the distinctions among politics, nature, art, and affect altogether, thereby creating a mindset that would ratify the war on dirt around a national "zeal."

"On the Great Western Canal of the State of New York" contradicts the politics of Freneau's earlier poem "The Projectors" in two ways. First, as the ode states, "Nature, herself, will change her face / And arts fond arms the world embrace" (Freneau 1993: 3). This literal sleight of hand—the poem imagines that nonhuman nature will somehow change its own face, rather than the shovel—reduces the distance between nature and art as they "embrace." With this poetic bridging, the poem can then install the Erie Canal as a "vast object" (4), a "work so vast" (5): that is, as sublime. And with this aesthetic transformation in place, further, this "work progress" (5) spans the gap between internal improvements and the liberal state: "Advancing through the wilderness, / A work, so recently began, Where Liberty enlightens man: / Her powerful voice, at length, awakes / Imprisoned seas and bounded lakes" (3). The Erie Canal attains its sublimity not only through its vastness but also through its liberating effects: endowed with agency, in cutting its path it unchains seas, unbinds lakes, and frees people from "tyrant[s]" (4). No longer seen as a threat to the yeoman farmer's freedom, internal improvements actually could effect his liberation.

Alongside reports by Hamilton and Albert Gallatin, imaginative literature thus aimed "the great idea [of infrastructure] to pursue" (Freneau: 1993: 3). Yet in nudging "practical action" (Hostetler 2011: 58), Freneau combined both spatial and temporal arguments in his anticipatory celebration of the canal so as to contain its meaning in advance. For a recent theorist of the temporality of infrastructure, Akhil Gupta (2018: 63), "we can understand a great deal about social futures by looking at infrastructure." In a telling overlap with Barlow's terminology from 1787, for Gupta, "Infrastructures are concrete instantiations of visions of the future": they are objects, in a line I cite in part above, "built in anticipation of a not-yet-achieved future." Perhaps even more so than Barlow's epic, Freneau's ode thus anticipates the "not-yet-achieved future" of the canal and thereby guarantees its aesthetic reception. After all, when it comes to infrastructure, the stakes are high: "We know that infrastructures fix space and time because, once finished, they are hard to reverse" (63).

If Freneau's poem collapses distinctions among politics, nature, and art, its temporal maneuverings further aim to link inversely the canal's construction timeline and its projection as an icon of American exceptionalism. Freneau's (1993: 4; emphases original) speaker crows, "To Fancy's view, what years must run, / What ages, till the task is done! / Even *truth*, severe would seem to say, / *One hundred years must pass away*." Although "the unrivalled work would endless seem," in reality, from the speaker's vantage point in 1820, "Three years elapsed, and behold it done! / A work from Nature's chaos won; / By hearts of oak and hands of toil / The Spade inverts the rugged soil" (5). The Erie Canal, its scale and its timely construction, proved wrong those "foreign nations" that deemed "our republic . . . incapable of works of great magnitude" (Colden 1825: 5).

And at the bottom of "this great enterprise" (5)? The humble "Spade," which effected the physical manifestations of the wild imperial surmise. Idealized by Freneau as humble but figured in the era's print culture as industrial (see fig. 2), as it altered the landscape, the shovel rested at the intersection of spatial and temporal axes and thus managed the "dynamic nature of infrastructural time" (Gupta 2018: 62) by uniting vision and reality, present and future. To put this argument slightly differently: the shovel could partner with romantic genius to

Figure 2 Earth-moving operations at the deep cut, near Lockport, New York, during the construction of the Erie Canal, an image included in Colden 1825. Smithsonian Institute Libraries

write the script of internal improvements. Adapting the wild surmise of infrastructure, in 1820 DeWitt Clinton captures this complementarity between the shovel and the imagination. Traveling the canal-in-progress near Ithaca, and writing under the name Hibernicus (1822: 9), he observes: "From Schenectady to the south end of Cayuga or Seneca Lakes, you may proceed by an uninterrupted navigation to the extent of near 250 miles—which will be enlarged when the canal reaches the Genesee river . . . Imagination, in this case, lags behind reality, and the utmost stretch of poetic vision becomes embodied into existence." Here, alongside the ditch-diggers' progress, Clinton records the poem immanent in the landscape—a strenuous act of "poetic vision" that nearly upturns the ground itself and guarantees the not-yet-completed romance of infrastructure in a triumphalism that blunts the fact that such projects are "always in process, always shifting, changing, decaying, being rebuilt, and being maintained" (Gupta 2018: 74).

Enlisted to stabilize these projects' symbolic meanings over time, such poetics concomitantly engaged "how these works, while being constructed, displace millions of residents in order to redistribute resources to a relatively more powerful few" (Appel, Anand, and Gupta

2018: 10). If Freneau's (1993: 3) ode views this mission as "true to honor's cause," so too does Cadwallader D. Colden's *Memoir*, "the written testimony that commemorated" (Seelye 1985: 241) the Grand Canal celebration in 1825. There Colden (1825: 178) speaks to infrastructure's "benign effects," as his text euphemistically has it, and reveals how infrastructure aimed to "differentiate populations and subjects" (Appel, Anand, and Gupta 2018: 5) along racial lines. The text imagines "How different will be the scene now presented" than when Henry Hudson first settled New York (Colden 1825: 4): "Instead of savages in their canoes . . . , there will be magnificent barques . . . bearing thousands of our fellow citizens, exulting in the accomplishment of a work which is an evidence, how immeasurably civilized, transcends savage man." In this vision of transcendence, Colden likely had in mind the Seneca Nation, the westernmost member of the Six Nations. The most "magnificent barque" in the "The Aquatic Display" (Colden 1825: 122) commemorating the completion of the canal, the ship that carried not only Clinton from Lake Erie to the Hudson River but also a keg of lake water to be mingled with the Atlantic, was named the *Seneca Chief*. The "Display" "transcended" all anticipation (122), and as Clinton wed the waters of "Lake Erie with the Atlantic," he prayed that "God . . . smile most propitiously on this work, and render it subservient to the best interests of the human race" (271).

Anticipating the Jacksonian ethics of infrastructure, for Clinton, the "best interests of the human race" included expropriating land for "our fellow citizens" where previously, in Freneau's (1993: 5) orientalist idiom, "tigers ranged and *Mohawks* trod." This is the infrastructural variant of a process described by Jean O'Brien (2010) of separating American Indians ("savage man") from modernity by consigning them to some remote past "before" infrastructure. In the modern settlerscape, ships laden with "products that new regions boast" (Freneau 1993: 4) replace an ancient Other, evidencing "the process of replacement [that] maintains the refractory imprint of the native counter-claim" to the land (Wolfe 2006: 389). In the politico-aesthetics of internal improvements, the poet-politicians integrate indigeneity into their sublime politics by replacing it. This dovetails with romantic texts that melancholically recur to "vanishing American" discourse so as, in Jillian Sayre's words (2018: 724), to "confirm the presence and sovereignty of the nation in the foreclosure of native futurity." Just as much as a sketch or a historical romance, then, an ode to a canal "imprints" in order to efface Indigenous claims to the settlerscape.

If, as Laurence M. Hauptman (2001: 3) has argued, internal improvements in western New York "led to the undoing of the Iroquois," the costs of their violence could be tallied in ecological figures as well. As W. B. Langbein (1976: 49) concludes for the US Geological Survey, the Erie Canal Commissioners of 1811, who surveyed, reported, and planned the canal, "were apprehensive and assertive that the resulting land development would have adverse environmental impacts." The canal commission's concern over these impacts—they especially worried over the canal's water supply—was not for the environment itself but for how they would affect farms and "the progress of industry" (Canal Commissioners of 1811 cited in Langbein 1976: 49). Understanding its project as an extension of what Kyle Powys Whyte terms an "industrial settler campaign," which captures the notion of settlement as a "sustained, strategic, and militaristic" "homeland-inscribing process" (Whyte 2017b: 208), "the operational view of the canal was economic and not ecologic" (Langbein 1976: 57).

Freneau's (1993: 5) ode asks: "To make the purpose all complete / . . . What rocks must yield, what forests fall?" Anticipating the cultural byproducts of the Grand Canal Celebration (Colden 1825) further, Freneau (1993: 5) adapts classical mythology to justify this assault: "With patient step I see them move / O'er many a plain, through many a grove; / Herculean strength disdains the sod / Where tigers ranged or Mohawks trod; The powers that can the soil subdue / Will see the mighty project through." Here nonhuman nature becomes a combatant that impedes Art's triumph. But this is not a fair fight. The shovel that "inverts" also "subdue[s]": it is the implement not of the common laborer but of "Ye artists, who, with skillful hand, / Conduct such rivers through the land" (5). The digging implement qua writing instrument thus sought aesthetically to transform dirt into poems, Native Americans into boats.

As Freneau navigated the "dynamic nature of infrastructural time" in the midst of the canal's construction (Gupta 2018: 62), the official poet of the Grand Canal Celebration, Samuel Woodworth, faced a different problem: now that the imagined future was here—now that the dirt was flung and Native Americans were replaced—how to turn it into what Jessica Hurley (2017: 771; emphasis original), focusing on postmodern infrastructures of waste, has called an "ongoing *now*"? The solution was to imbue the canal with a politico-aesthetics that extended the present indefinitely, and Woodworth's (1825: 252) poem, "Ode for the Canal Celebration," imagines infrastructure as a means of ever reproducing the settler state—the canal a grand work "That will glow through future ages, / And cover with glory and endless fame / Columbia's immortal sages." There are no shovels in the verse, but the "deathless" significance of those martial "strains" that flung the dirt is clear: they "shall unborn millions yet awake" (253). Built in "anticipation of a not-yet-achieved future," the Erie Canal figures here as a "technology of and for the future" (Gupta 2018: 63). The canal is thus a device that initially imagined but now delimits that future . . . forever: "'tis done," but not quite, as this sublime icon of the nation-state "shall perpetuate / The glory of our native State" (Woodworth 1825: 252).

Hawthorne's "The Canal-Boat": The Prose of Path Dependence

In seeing art to "teach young Genius to rise from earth," Woodworth's (1825: 252) "Ode for the Canal Celebration" imagined internal improvements extending what Freneau (1993: 4) termed "A *new Republic* in the west / (A great example to the rest)." Writing from the antebellum era, however, reveals the flipside of the infrastructural dialectic, in which some writers imagined canals and canal travel instead as dull, irritating, and enervating. As early as 1811, the Erie Canal Commissioners were aware of the threat of "the early obsolescence of the canal" (Langbein 1976: 10) by steam-powered technology. Harriet Beecher Stowe's "The Canal Boat," first published in *Godey's Lady's Book* in 1841, attests to the velocity of this outcome but couches her invidious claims about a horse-drawn canal boat not only in technological but in aesthetic terms as well. Whereas "there is something picturesque, nay, almost sublime in the lordly march of your well built, high bred steamboat," the canal boat is one of "the most absolutely prosaic and inglorious" "ways of traveling" (Stowe 2003: 96). Leo Marx, in his classic study *The Machine in the Garden* (1964), shows how in the 1840s the steamboat became the latest icon of "the sublime progress of the race" (Marx 2000: 197). For Stowe (2003: 97), the newer technology evinced the dynamical sublime: "Then there is something mysterious, even awful, in the power of steam," and the ship has "claims both to the beautiful and the terrible." To the contrary, "in a canal boat," one finds "there is no power, no mystery, no danger; one cannot blow up, one cannot be drowned, unless by some special effort: one sees clearly all there is in the case—a horse, a rope, and a muddy strip of water—and that is all" (97). Whereas Freneau and Woodworth layered multiple meanings onto artificial rivers, Stowe would labor in vain: how to poeticize what "one sees clearly"? Rubenstein, Robbins, and Beal (2015: 576) discuss how infrastructure can elude analysis, since it "tends to go unnoticed when it's in fine working order"; "destroying it [such as often happens in action movies] is simply one strategy for making it appear." The threat of its annihilation unthinkable for Stowe, in 1841 there is no longer any uncertainty of meaning in need of the poet's management. There is no catalogue verse of canal travel's sublimity; instead, there is a "prosaic" "catalogue of distresses" (Stowe 2003: 102).

This perceived aesthetic poverty of canal travel reveals the contradictions of the infrastructural dialectic set in motion by an ongoing industrial settler campaign. In contrast to Washington's wild surmise in the upstate forest, Stowe's (2003: 102) "catalogue" must be foreshortened: "our shortening paper," she writes, "warns us not to prolong our catalogue of distresses beyond reasonable bounds." Whereas Washington's gaze constructed the bounds of a "new empire," for Stowe, to linger here would only spoil "the invisible, forgettable ambiance in which the daily drama of modern life takes place" (Rubenstein, Robbins, and Beal 2015: 585). She concludes her sketch not with a generalization to avoid canal travel, nor even with a surprising mandate to embrace its sublime potential. Rather, she advises her reader "to take a good stock both of patience and clean towels" for one's journey (Stowe 2003: 102). After raising canal boats from their "inglorious" lowness so everyone "sees [them] clearly," she restores them to invisibility so as to carry on the quiet drama of daily life.

Just as Stowe traveled in canal boats across the American West in the 1830s, so too did Hawthorne.[14] Like Stowe's work, his sketch "The Canal-Boat," first published in 1835 in a series for the *New-England Magazine* called "Sketches from Memory by a Pedestrian" and later included in his second edition of *Mosses from an Old Manse* (1854), casts canal travel as more akin to the stuplime than the sublime. There Hawthorne (1974: 437) strikes down the Erie Canal's claims to glory and engages the processes that had enabled this retrograde travel experience in the first place: the processes that in altering the landscape enthroned the "savage queen [Nature] . . . on the ruins of her empire." But in tallying the ecological, biopolitical, and aesthetic costs of internal improvements rationalized by Freneau and others, Hawthorne offers an even starker failure than Stowe does to disrupt the relation between aesthetics and empire: a failure to halt the infrastructural dialectic by leaving the cut path in search of a more just one.

Just as Thomas Cole visualized a cyclical philosophy of history in his *The Course of Empire* series (1833–1836), Hawthorne's infrastructural work reveals the flip side of "improvements"—ruins—as a necessary corollary and recasts the desiderata of the era's sublime politics as liabilities, rather than as accruing assets effected by the distribution of resources, flow of people, and alterations to the landscape. The sketch begins with the narrator "inclined to be poetical about the Grand Canal"—the "new river" (Hawthorne 1974: 429). Referring to the classicism of the Grand Canal Celebration from 1825, he pictures his canal boat and its "three horses harnessed to our vessel, like the steeds of Neptune to a huge scallop-shell, in mythological pictures" (430). And with the commercial triumph of the canal over the "rage of war" in mind, he wants to see the canal as a sublime object that comes to fruition through human Art's devotion to the script of progress: "In my imagination," he writes, "De Witt Clinton was an enchanter, who had waved his magic wand from the Hudson to Lake Erie, and united them by a watery highway, crowded with the commerce of two worlds, till then inaccessible to the other" (430).

This expectation for the sublime quickly cedes to stultifying boredom, and Hawthorne's sketch drags the reader down into the mud rather than into the "uplifting transcendence" of the sublime (Ngai 2005: 267). The canal, in Ngai's words, "obstructs aesthetic or critical response" (262). In her theorization of the stuplime, Ngai sweeps aside Romantic texts, beholden to a popularized Kantian sublime, because they "tend to emphasize the self's initial feeling of limitation or disempowerment and thus to formulate the sublime primarily as an experience of being astonished and overwhelmed by a vast or intimidating object" (267). To the contrary, stuplime works "tend to draw us *down* into the sensual and material domain of language and its dulling and irritating iterability, rather than elevating us to a transcendent, supersensible, or spiritual plane."

Yet Hawthorne drags his characters not just through the "'common muck' of language" (Ngai 2005: 268) but also through the literal mud of the canal. For Hawthorne, though, this is a compulsion for neither mastery of (as in Washington) nor intimacy with (as in Walt Whitman's "Song of Myself" [1855]) the landscape. Rather, the mucky language that pervades the sketch's discourse reveals infrastructure as an impossible site to reconstitute the (Romantic) self. The water is an "interminable mud-puddle" (Hawthorne 1974: 430); the boat is "dirty" (432), so, too, is the water that holds the "foolish birds" that the passengers "pelt . . . with apples" (433); and the view, such that it is, reveals "dismal swamps and unimpressive scenery" (430). Preventing the elevation required for transcendence on this "dead flat between Utica and Syracuse" (436), the canal cannot "properly mobilize" the sublime (Ngai 2005: 270).

Instead, its dirtiness activates in Hawthorne *something much closer to an ordinary fatigue*" (Ngai 2005: 270; emphases original). Put otherwise, Hawthorne's (1974: 434) "intolerable dullness" calls attention (in Lauren Berlant's elegant phrase) to "the wreck of the old good life fantasy" (Berlant 2016: 398). In response to the iterative, dulling scene of a voyage, the sketch bears witness to wrecks nevertheless. In this vein, it offers a pattern of falling in response to the boat's "tiresome . . . reality" (Hawthorne 1974: 433). And in a single paragraph, we see that "the tow-rope caught a Massachusetts farmer by the leg, and threw him down in a very indescribable posture"; "a new passenger fell flat on his back, in attempting to step on deck"; and "another, in his Sunday clothes, as good luck would have it . . . forthwith plunged up to his third waistcoat button in the canal, and was fished out in a very pitiable plight" (433).

An iterative performance that shows how "stuplimity drags us downward . . . rather than transporting us upward toward an unrepresentable divine" (Ngai 2005: 273), this pattern of falling might have led to a "resistant stance" (297) to a vast but finite system inscribing the settlerscape—a stuplime politics that pantomimes ironically the boosterism of the canal. As Hawthorne (1974: 430) realizes early in his sketch, this "most fertilizing of all fluids," the "new river," has been selective in its resource distribution. He couches this realization in class terms—he sees a woman who "looked like Poverty personified" (432) alongside the canal bank. But countering the belligerence of Freneau's ode, Hawthorne also conveys his critique in racial and ecological terms and thus glimpses an anticolonial reading of the Erie Canal.

Perhaps in dialogue with Colden's comment regarding American Indians "in their canoes," Hawthorne's sketch actually countenances one such craft: contrary to the usual "line boats" or "light packets," he "encountered a boat, of rude construction, painted in all gloomy black, and manned by three Indians, who gazed at us in silence and with a singular fixedness of eye" (431). As a whole, the sketch includes several moments of shared gazes among the passengers, but this gaze, the lone Indigenous presence in the sketch, stands out as an example of people on the canal boat being seen from without. One imagines that, contra Stowe, these Indigenous people see more in the ship than just a rope and a boat. But for Hawthorne, too, this experience, which so briefly dwells in the possibility that "perhaps, these three alone . . . had attempted to derive benefit from the white man's mighty projects, and float along the current of his enterprise" (431), also sticks: a sign that he witnesses—and feels seen as a participant in—the actual and the epistemic violence of infrastructure's creation.

The normalization of this violence, reinforced by early Republican writers as a "benign process involving righteous relations with Indians and just property transactions" (O'Brien 2010: xv), begins to fray in Hawthorne's sketch. Berlant's theorization of the infrastructural glitch can clarify here. This "singular fixedness of eye" (Hawthorne 1974: 431) suggests "the glitch of the present as a revelation of what *had been* the lived ordinary" (Berlant 2016: 403; emphases original). Whereas Colden wants to erase the Indigenous presence from the canal, and whereas another literary tradition represented by William Cullen Bryant's "The Prairies" (1832) wants to excise Indigenous people's historic relation to the manipulated earth entirely, Hawthorne (1974: 431) imagines a past "lived ordinary" in which the Native Americans were "ancient possessors of the land."

Yet even with this attempt to illuminate Native Americans who were adversely affected by the canal, Hawthorne cannot sever ties to coloniality. For here he recurs to an ideology that denigrates Indigenous people as incapable of adapting "to the changes wrought by colonialism by selectively embracing new ways and ideas" (O'Brien 2010: xxii). In referring to Indigenous people as "ancient," Hawthorne uses a construct that "served to purify the landscape" of their presence (O'Brien 2010: xxii). In this case, Hawthorne performs contradictory cultural work typical of the historical settler narratives that O'Brien studies: on the one hand, his sketch explicates infrastructure's temporality as part of an imperial process that inflicts injustice in the present. On the other one, he turns away from this scene and, in O'Brien's term, "lasts" "these three alone" as if to diminish the harmful effects of canal development on Indigenous societies.

A microcosm of the more general contradiction within the dialectic of infrastructure—the aesthetic swing from sublime (or romantic) to dull (or invisible)—this take on the "ongoing *now*" of infrastructural violence tracks along an ecological axis as well. Though "The Canal-Boat" devotes only two sentences to an Indigenous presence, it expatiates on the "riotous destruction" of the forest (Hawthorne 1974: 437). Rather than envisioning the possibilities of infrastructure "o'er the pathless main," Hawthorne gazes instead upon the "diseased splendor" of the "dead forest" (438). "In other lands," he quips, "decay sits among fallen palaces; but here, her home is in the forests" (437). Envisioning a dead nonhuman nature from a deadening ship, he thus turns toward "another emblem" (437) of infrastructure's "rage of war." If infrastructure does "not appear to be doing anything" (Wakefield 2018: 2), Hawthorne (1974: 436) makes clear that it already has done a lot—the "forest is now decayed and death-struck."

What Hawthorne presents as an "emblem" of the American technological *stuplime*—the "death-struck" forest—was the product of preparations to enhance the land's surplus value. The canal commissioners of 1811 worried, "in the progress of industry," whether deforestation would evaporate the water supply for the canal and the "ponds . . . collected for mills and other machinery" (quoted in Langbein 1976: 49). (It would not.) By 1824, as worries over toll receipts that would fund the canal debt intensified, the commissioners turned their attention to "the canal's climactic handicap— winter ice" (Langbein 1976: 54). Through "changes of climate . . . by cutting down of the forest," they hoped, "our annual seasons of navigation will ultimately be extended to 250 or 275 days" (quoted in Langbein 1976: 53). Although the average season would actually last 220 days, the implication is clear: the commissioners pursued "climate destabilization" that would "dramatically change ecosystems" and "obstruct indigenous peoples' capacities to adapt to the changes" (Whyte 2017b: 209). As an emblem of the industrial settler campaign, the "death-struck forest" of the canal thus connects a disrupted Indigenous ecology to the dystopic "Anthropocene futures" that follow from the ongoing processes of coloniality.

In response to these disruptions, Hawthorne's "The Canal-Boat" is finally stupefied not only by the fixed stare of the Native Americans but also by these "dismal black stumps" (Hawthorne 1974: 432)—glitches of "resource distribution, social relation, and affective continuity" (Berlant 2016: 394) that emanate from the construction of the Erie Canal into the present.[15] There is potential for revolution within the stuplime. Stuplimity, Ngai (2005: 262) writes, "prompt[s] us to look for new strategies of affective engagement and to extend the circumstance under which engagement becomes possible." And if Ngai's aesthetic theory suggests how to brighten Hawthorne's dismal words, for Berlant's (2016: 394) infrastructuralism, these glitches potentially alert us to a "critical social form" that might "alter . . . infrastructures of sociality itself." Or, to put this otherwise, Hawthorne's text might have enacted the desiderata of a critical infrastructuralism.

In a recent review essay, Stephanie Wakefield (2018: 8) finds infrastructuralism to be a field "that so many are now studying," as the "old modes of thinking and acting are coming undone." Perhaps this growth attests to interest in the possibilities that arise from the ruins of modern infrastructure, as ruins discourse confronts life after empire's decay. Yet, a failed stuplime politics, "The Canal-Boat" reveals the power of path dependency, infrastructuralists' term to capture the "strange staying power" of institutions and their praxes (Levine 2010: para. 71–72), and ends by abandoning the crises of the canal rather than imagining a "reconciliation" between "indigenous people and members of settler society to learn about how humans are entangled with other species and with the environment" (Whyte 2017b: 213).

When seen through the lens of critical infrastructuralism, this abandonment seems even starker since it is formulated as a reaction to a technical glitch on which infrastructure studies hang their vision of reconciliation. In the middle of the night, the same towrope that had previously thrown down a Massachusetts farmer "got entangled in a fallen branch on the edge of the canal, and caused a momentary delay" (Hawthorne 1974: 437–38). During the stoppage, Hawthorne meanders off the path "to examine the phosphoric light of an odd tree, a little within the forest" (438). What he finds is a "frigid fire, a funeral light, illumining death and decay." One might offer a new materialist reading here of the posthumous agency of the lively world— the fallen limb that exposes the canal's "moral rottenness." Yet incapable of resisting the necropolitics of the shovel, as he "recollected" himself, he finds the boat has left him behind—*and he is ecstatic*. Released from the waste of the canal, he exclaims, "They are gone! Heaven be praised!" This "ejacula[tion]" (438) relieves the narrator from the ongoing crises of infrastructure that he would rather repress than engage: its vistas of the unequal distribution of wealth, the forced relocation of Indigenous people, and of course the war on the environment. Instead, he embarks on a "comfortable walk" and follows the cut path to the nearest inn. And to light his way in the dark? Hawthorne takes a "flambeau from the old tree, burning, but consuming not" (438). Lighted by this bit of living death that burns but consumes not, he hews to the path, certain his property will be restored to him at last.

Coda: Digging Side by Side

If the built environment of upstate New York can be read as an imperial poem—be it sublime or stuplime—perhaps no settlerscape has a longer relation to coloniality than the Panama Canal. Yet, like the Erie Canal, the physical site that joins the Atlantic and Pacific Oceans is equally a textual one of coloniality's wild surmises: a site from which to both read and write the war on dirt, running from July 4, 1817, when a canal contractor, Judge John Richardson, "drove his spade into the ground" of New York City (Sheriff 1996: 9) to November 16, 1906, when Theodore Roosevelt operated a Bucyrus steam shovel in Colón, Panama (see fig. 3). In excavating the Culebra Cut, this machine, operated solely by white men in the strictly segregated Panama Canal Zone,[16]

Figure 3 President Roosevelt running an American steam shovel at Culebra Cut, Panama Canal, circa November 26, 1906. Library of Congress

would overcome the "dread Isthmus" and build "New paths" (Barlow 1787: 246). In operating this machine, Roosevelt thus fulfilled the imperialism immanent to nineteenth-century internal improvements in a vertiginous spatio-temporality connecting Columbus (Colón) and Hernán Cortés to the US presidency.

That Roosevelt renews the infrastructural dialectic by hailing laborers as imperial soldiers makes his discourse stand out in the works I have tracked in this essay. The "common laborers whose physical strength made the [Erie] Canal possible," as one historian writes, were "virtually" excised from the "hundreds of speeches and toasts at the official gatherings" to celebrate its completion (Sheriff 1996: 46)— and, one might add, from Stowe's and Hawthorne's travel sketches, which feature magic wands, not shovels. Undoing this erasure of labor is central to Margarita Engle's *Silver People*, a retelling in verse— targeted to younger readers—of the story of the Panama Canal's construction. As her character, Panama Canal geologist, Augusto, puts it in his account of the Panama–Pacific International Exposition in San Francisco, "There are no monuments to the tens of thousands who died in Serpent Cut mud. No one cares. No one cares because no one knows" (Engle 2014: 230). If one feature of modern infrastructure is that one cannot contemplate its sublimity lest one's purpose is derailed, that amnesia seems to erase labor from the field of vision as well.

The politico-aesthetic project of critical infrastructuralism opposes this amnesia and pursues a project of defamiliarization that raises a "host of urgent questions about access and ownership, rights and responsibilities" (Rubenstein, Robbins, and Beal 2015: 585). To overcome the infrastructural dialectic, infrastructuralists view works as a "sprawling sociotechnical and environmental system" that includes "laborers, engineers, hydrologists, accountants, and much more" (Carse and Keiner 2016: 208). And this approach to infrastructure as an unwieldy assemblage is precisely the tack that Engle takes in *Silver People*. To redeem the failed stuplime politics of writers like Stowe and Hawthorne, as well as early twentieth-century poets of the canal,[17] she fleshes out what the historian Julie Greene (2016: 4–5) has termed the "imperial labor migrations" to the Canal Zone that helped extend the US empire "from the ground up." Aestheticizing the "sprawling" nature of infrastructure, Engle's (2014: 68) novel gives voice to fictional personages, including Indigenous Panamanians ("here long before the Americans, / before the French, / even before the bold adventurers"), Caribbean laborers, and Hispanic scientists; historical ones, including Roosevelt, General George Washington Goethals (chief engineer of the canal), and Gertrude Beeks (a health inspector and secretary of the Welfare Department of the National Civic Federation); as well as indigenous nonhuman species of flora and fauna, including frogs, birds, and trees. In doing so, its politico-aesthetics stand where Hawthorne can only fall down in the muck, and the novel realizes a stuplime politics, as it anticipates the potential for a transformative life among ruins.

The narrative accomplishes this multivalence with a diegesis consisting of species affected by—and that help effect—infrastructure. Such a technique stands in contrast with the univocality of the early Republican and Jacksonian texts examined above. Whereas Hawthorne's "three Indians" are shrouded in silence, Engle's (2014: 68) heroine, an Indigenous Panamanian healer, Anita, sparkles with agency: she "row[s] from village to village / in an old dugout canoe" during the rainy season because "she can't let something as common as water / keep me from working to help sick people." Anita's counterpart, furthermore, a Cuban migrant named Mateo drawn to the Canal Zone so as to escape his violent father and strive for improved economic prospects,[18] develops an ecological consciousness as he digs the earth—he mourns that the Serpent Cut connecting Gatun Lake to the Atlantic Ocean "just a few years ago / . . . was still a forested / mountain" (106). And contra Hawthorne's "dead forest," the trees insist in their own narration that they "Are here / Always here / Always" (35) but acknowledge that "men / with machines and explosives have made some of us vanish" (67).

Such multivalence enables Engle to resist "the theme of the techno-scientific conquest of nature" typical even of recent revisionist historiographical accounts of the Panama Canal (Carse and Keiner 2016: 213). To break the infrastructural dialectic, her novel models instead "ruins thinking" (Wakefield 2018: 9): a discourse that imagines in the canal the potential for (in a passage from Berlant cited above) a "critical social form" that might "alter . . . infrastructures of sociality itself" (Berlant 2016: 394). The characters, who portray "the history of empire from the perspective of those who built it" (Greene 2016: 14), sense "a society that is ready to reinvent the world" (Wakefield 2018: 8) after the ecological devastation that the canal occasioned, sustained, and would continue in its facilitation of global trade. Engle (2014: 213–14) marks this orientation with a section called "Sky Castles 1914" set "when all the digging finally ended" and the "forest" became a lake, as Mateo and Anita bear witness to infrastructure's planned violence. Just as they "dig side by side" (212), they narrate together the consequences of this human-altered hydrology: "crate towns vanished, and at least fifty thousand silver people [racialized laborers paid in silver, rather than gold] / had to flee for their lives / along with native villagers, / and all the wildlife" (214). What remains is "this one mountain peak": "It's an island now—our new home . . . we think of as our sky castle" (214–15).

What remains is a platform from which to reject the appetitive, imperial wild surmises that run through the infrastructures of coloniality from the past to the present. Unveiling his infrastructure initiative in 2018, Donald Trump grounded his plan to invest $1.5 trillion in the United States' past achievements. "We dug out the Panama

Canal," he told those assembled just south of Lake Erie in suburban Cleveland, Ohio, at the Local 18 excavation training site. Attesting to the compulsion that the wild surmise of infrastructure entails, he continued: "We must recapture the excitement of creation, the spirit of innovation, and the spark of invention" that will allow the United States to "reclaim that proud heritage." "We will tear down every obstacle," Trump promised, and "shap[e] our destiny." In a jeremiad that aimed to reanimate the surmise of Keats and Barlow, Trump clearly saw infrastructure as a site of sublime politics.

For Anita, Mateo, and their Jamaican friend Harry, however, they stand "Silent, upon a peak in Darien" (Keats 1978: 64), primed to enact a critical infrastructuralism rather than a reclaimed "Powerful American pride" (Trump 2018). To borrow Wakefield's (2018: 9) words: "they are using the experimental audacity of resilience and the 'here and now' mentality of ruins thinking to create their own ways of living irreducible to liberal life" in order to decouple the wild surmise and coloniality from world-building. Whereas the early Republican poets anticipate infrastructure by containing its violence as "improvement," Engle's characters build their future from the ruins of this process. Their surmise atop the mountain-island is as anticipatory as the imperial surmise, and it is as "silent." But what distinguishes it is its "imagining a livable provisional life" (Berlant 2016: 395): "Our only way of speaking / with the forest / is silence. / We watch. We study. We record / all that we see as we peer / into humming bird nests / and howler eyes. / We listen . . . and hope to remember every detail of beauty" (Engle 2014: 220).

And in this attentive silence, it is they who are transformed by the already transformed settlerscape: "Sometimes we feel like strangers, / and other times, we feel / transformed / into a natural part / of this wild / world" (221). As one-time diggers, having helped to expand where "empires lay," and, as an herbal healer, having ministered to the fallen in the "rage of war," these young people acknowledge the Panama Canal as part of coloniality's ongoing planned violence. In doing so, they refuse the break that modern infrastructure enacts between ecological and social resilience (Whyte 2018: 125). Rather, they commit to the "single goal" of "learning to understand / this one mountain-island" (Engle 2014: 223). They do so neither to "return to a 'before' modern infrastructure" (Wakefield 2018: 9) nor to reclaim a "proud heritage" (Trump 2018). Atop the settlerscape that would join the Erie Canal as another symbol of US domination of globalized trade—and thus one more emblem of the Anthropocene, as Davis and Todd might argue—they embrace the "provisional, uncertain, and unpredictable experiences of experimental practices" (Wakefield 2018: 9) actively refused by Barlow's and Freneau's sublimity and deemed impossible by Stowe's and Hawthorne's failed, path-dependent stuplime. Instead, in building "a sense of identity associated with the environment and a sense of responsibility to care for the environment" (Whyte 2018: 127), their humble surmises invite readers to anticipate a world "for now."

Andrew Kopec is associate professor of English at Purdue University Fort Wayne. His essays have appeared in *Early American Literature*, *ELH*, *ESQ*, and *PMLA*, in addition to other journals and edited volumes. He is finishing his first book, on the relation between American Romanticism and the boom-bust business cycle. The current essay draws from research for a new project focused on the trans-American infrastructures of the "transportation revolution."

Notes

For their support as I worked on this essay, I would like to thank the members of the Civil War Caucus and my colleagues in English and Linguistics at Purdue University Fort Wayne, Shannon Bischoff and Damian Fleming.

1 Barlow's later poem *The Columbiad* (1807) incorporates *The Vision of Columbus*. Following Wertheimer (1999: 74), who identifies the 1787 poem as "more compelling" than the 1807 one and as the "true source of the imperial Pan-American imaginary," I quote from *Vision*.

2 For more on infrastructure as "planned violence," see Rubenstein, Robbins, and Beal 2015: 580.

3 See also Wynter 2003.

4 Previous attempts at canal development, including the Patowmack Company, organized by George Washington in the late 1700s, had failed in the early United States. See Bernstein 2006: 22–23. See also Taylor 1951: 32–33.

5 In the context of the Panama Canal, Ashley Carse and Christine Keiner (2016: 214) develop similar thinking: "Seen from an environmental perspective, the year 1914 did not mark the end of the canal story, but one pivotal moment in an ongoing process of transformation from colonialism to the present."

6 Throughout this essay, I follow Ashley Carse's (2012: 543) view of infrastructure as more than "hardware." It is also a "revealing site for ethnographic research on negotiation, struggle, and meaning."

7 Here I follow D. Ezra Miller (2016: 210) who uses the term "settlerscape" "to draw attention to the way clearly recognizable Indigenous landscapes became transformed into distinctive, settled communities. [It] recognizes the competition and complementarity of interests converging in particular places, only this time during the settlement period." See also Whyte 2018: 136 and Carse 2012: 540.

8   In a demonstration of the axe as an implement of coloniality, Catharine Maria Sedgwick's historical romance *Hope Leslie* (1827) refers to an "English axe," which fells both "forest trees" and (in the Pequot character Nelema's term) "my race" (Sedgwick 1987: 37).

9   In an essay about "weather as a figure for the *reversibility* of foreground and background, everyday and event" in late-nineteenth-century US literary realism, Nathan Wolff (2018: 225) observes a related oscillation within "enduring questions about race, ecology, and history." My thanks to the author for his feedback on a late version of my project.

10   For Pease (1984: 278n9), the settler could "proclaim that only through destruction of Nature's bounty could he feel, by doing what Nature commanded, as if he were truly in touch with Nature's will." See also Wolfe 2006: 388.

11   In fact, in his poem "Sir Harry's Invitation" (1779), regarding British General Henry Clinton's treatment of Tory refugees, Freneau (1963b, 2:7) had associated "dig[ging] the canal" not with a republic but with a monarchy: for "the king" who "wants your aid."

12   In her counterintuitive reading, Elizabeth Hewitt (2019: 630) identifies Hamilton as a romancer of numbers, "gambling on possible risk, and trying to predict the possible outcomes that may follow from . . . individual choices."

13   Jefferson's dubiousness was well-founded. According to the historian George Rogers Taylor (1951: 32), "by 1816 only about 100 miles of canal had been constructed in the United States." "The building of the Erie Canal," Taylor concludes, "was an act of faith" (33).

14   Hawthorne focused on other infrastructures of the "transportation revolution" in works devoted to rails ("The Celestial Railroad" [1843] [see Marx 2000]), turnpikes ("Mr. Higginbotham's Catastrophe" [1834]), and drawbridges ("The Toll-Gatherer's Day" [1837]).

15   Even if the canal's construction did not realize the feared environmental impacts, the Erie Canal commission's instrumental view of the environment "probably led to neglect of consideration of environmental risks in subsequent public works practice during the 19th century" (Langbein 1976: 2).

16   As Roosevelt (1928: 13) stated in the Canal Zone regarding "the steam-shovel man," "he is the American who is setting the mark for the rest of you to live up to." This comment reveals one facet of the race-based hierarchy that was militantly policed in the Canal Zone. As the one-time Panama Canal Zone policeman and census enumerator Harry Franck says in Engle's (2014: 192) *Silver People*: "I was told that white men / should never be seen working / with shovels." See Greene 2016: 9.

17   For example, see John Hall's *Panama Roughneck Ballads* (1912), especially the poems "The Steam Shovel" and "The Price of Empire."

18   Poets, historians, and writers have documented the economic reasons that motivated trans-Caribbean people to dig the Panama Canal. See Senior 1977 and Greene 2016: 8.

References

Appel, Hannah, Nikhil Anand, and Akhil Gupta. 2018. "Introduction: Temporality, Politics, and the Promise of Infrastructure." In *The Promise of Infrastructure*, edited by Nikhil Anand, Akhil Gupta, and Hannah Appel, 1–38. Durham, NC: Duke Univ. Press.

Barlow, Joel. 1787. *The Vision of Columbus: A Poem in Nine Books*. Hartford, CT: Hudson and Goodwin.

Berlant, Lauren. 2016. "The Commons: Infrastructures for Troubling Times." *Environment and Planning D: Society and Space* 34, no. 3: 393–419.

Bernstein, Peter L. 2006. *Wedding of the Waters: The Erie Canal and the Making of a Great Nation*. New York: Norton.

Carse, Ashley. 2012. "Nature as Infrastructure: Making and Managing the Panama Canal Watershed." *Social Studies of Science* 42, no. 4: 539–63.

Carse, Ashley and Christine Keiner. 2016. "Forum Introduction" ("Panama Canal Forum: From the Conquest of Nature to the Construction of New Ecologies"). *Environmental History* 21, no. 2: 207–21.

Colden, Cadwallader D. 1825. *Memoir Prepared at the Request of a Committee of the Common Council of the City of New York, and Presented to the Mayor of the City, at the Celebration of the Completion of the New York Canals*. New York: Corporation of New York.

Cooper, James Fenimore. (1847) 1863. *The Crater*. New York: James G. Gregory.

Davis, Heather, and Zoe Todd. 2017. "On the Importance of a Date, or Decolonizing the Anthropocene." *ACME: An International Journal for Critical Geographies* 16, no. 4: 761–80.

Engle, Margarita. 2014. *Silver People: Voices from the Panama Canal*. New York: Houghton Mifflin Harcourt.

Freneau, Philip. 1963a. "The Projectors." In vol. 2 of *The Poems of Philip Freneau: Poet of the American Revolution*, edited by Fred Lewis Pattee, 160–61. 3 vols. New York: Russell.

Freneau, Philip. 1963b. "Sir Harry's Invitation." In vol. 2 of *The Poems of Philip Freneau: Poet of the American Revolution*, edited by Fred Lewis Pattee, 7–8. 3 vols. New York: Russell.

Freneau, Philip. (1822) 1993. "On the Great Western Canal of the State of New York." In *Philip Freneau to Walt Whitman*, vol. 1 of *American Poetry: The Nineteenth Century*, 3–5. Edited by John Hollander. 2 vols. New York: Routledge.

Gallatin, Albert. (1808) 1968. *Report of the Secretary of the Treasury on the Subject of Public Roads and Canals*. Reprints of Economic Classics. New York: Augustus M. Kelley.

Greene, Julie. 2016. Presidential Address: "Movable Empire: Labor, Migration, and U.S. Global Power during the Gilded Age and Progressive Era." *The Journal of the Gilded Age and Progressive Era* 15, no. 1: 4–20.

Gupta, Akhil. 2018. "The Future in Ruins: Thoughts on the Temporality of Infrastructure." In Anand, Gupta, and Appel 2018: 62–79.

Hamilton, Alexander. (1791) 1827. *Alexander Hamilton's Report on the Subject of Manufactures*. Edited by Mathew Carey. Sixth edition. Philadelphia.

Hauptman, Laurence M. (1999) 2001. *Conspiracy of Interests: Iroquois Dispos-session and the Rise of New York State.* Syracuse, NY: Syracuse Univ. Press.

Hawthorne, Nathaniel. (1835) 1974. "The Canal-Boat." In *Mosses from an Old Manse,* vol. 10 of *The Centenary Edition of the Works of Nathaniel Hawthorne,* 429–38. Edited by William Charvat, Roy Harvey Pearce, Claude M. Simpson, et al. 23 vols. Columbus: Ohio State Univ. Press.

Hewitt, Elizabeth. 2019. "Romanticism of Numbers: Hamilton, Jefferson, and the Sublime." In "Economics and American Literary Studies in the New Gilded Age," edited by Matt Seybold. Special issue, *American Literary History* 31, no. 4: 619–38.

Hibernicus [DeWitt Clinton]. 1822. *Letters on the Natural History and Internal Resources of the State of New-York.* New York: Bliss and White.

Hostetler, Michael J. 2011. "The Early American Quest for Internal Improvements: Distance and Debate." *Rhetorica* 29, no. 1: 53–75.

Hulbert, Archer B. 1920. *The Paths of Inland Commerce: A Chronicle of Trail, Road, and Waterway.* New Haven, CT: Yale Univ. Press.

Hurley, Jessica. 2017. "Impossible Futures: Fictions of Risk in the *Longue Durée.*" *American Literature* 89, no. 4: 762–89.

Jackson, Andrew. 1837. "Second Annual Message. December 7, 1830." In *Messages of Gen. Andrew Jackson: With a Short Sketch of His Life,* 85–126. Concord, NH: Brown and White; and Boston: Otis.

Kaplan, Amy, and Donald E. Pease, eds. 1993. *Cultures of United States Imperialism.* Durham, NC: Duke Univ. Press.

Keats, John. (1817) 1978. "On First Looking into Chapman's Homer." In *The Poems of John Keats,* edited by Jack Stillinger, 64. Cambridge, MA: Belknap Press of Harvard Univ. Press.

Langbein, W. B. 1976. *Hydrology and Environmental Aspects of Erie Canal (1817–1899).* Washington, DC: United States Government Printing Office.

Larson, John Lauritz. 1987. "'Bind the Republic Together': The National Union and the Struggle for Internal Improvements." *Journal of American History* 74, no. 2: 363–87.

Levine, Caroline. 2010. "Infrastructuralism; or, the Tempo of Institutions." In *On Periodization: Selected Essays from the English Institute,* edited by Virginia Jackson, para. 53–123. ACLS Humanities E-Book. Cambridge, MA: The English Institute.

Livingood, James W. 1941. "The Canalization of the Lower Susquehanna." *Pennsylvania History: A Journal of Mid-Atlantic Studies* 8, no. 2: 131–47.

Markwyn, Abigail M. 2016. "Queen of the Joy Zone Meets Hercules: Gendering Imperial California at the Panama-Pacific International Exposition." *The Western Historical Quarterly* 47: 51–72.

Marx, Leo. (1964) 2000. *The Machine in the Garden: Technology and the Pastoral Ideal in America.* New York: Oxford Univ. Press.

Mignolo, Walter D. 2007. "Introduction: Coloniality of Power and De-Colonial Thinking." In "Globalization and the De-Colonial Option." Special issue, *Cultural Studies* 21, nos. 2–3: 155–67.

Miller, D. Ezra. 2016. "'But It Is Nothing Except Woods': Anabaptists, Ambitions, and Northern Indiana Settlerscape, 1830–1841." In *Rooted and Grounded: Essays on Land and Christian Discipleship*, edited by Ryan D. Harker and Janeen Bertsche Johnson, 208–17. Eugene, OR: Pickwick.

Ngai, Sianne. 2005. *Ugly Feelings*. Cambridge, MA: Harvard Univ. Press.

Nye, David E. 1994. *American Technological Sublime*. Cambridge, MA: MIT Press.

O'Brien, Jean. 2010. *Firsting and Lasting: Writing Indians out of Existence in New England*. Minneapolis: Univ. of Minnesota Press.

Pease, Donald E. 1984. "Sublime Politics." In "On Humanism and the University I: The Discourse of Humanism." Special issue, *boundary 2* 12, no. 3–13, no. 1: 259–79.

Roosevelt, Theodore. (1906) 1928. "Extracts from Address of President Roosevelt to Canal Employees at Colon, November 17, 1906." In *The Canal Diggers in Panama 1904 to 1928*, 12–14. Balboa Heights, Canal Zone, Panama: The Panama Canal Retirement Association.

Rubenstein, Michael, Bruce Robbins, and Sophia Beal. 2015. "Infrastructuralism: An Introduction." In "Infrastructuralism." Special issue, *MFS: Modern Fiction Studies* 61, no. 4: 575–86.

Sayre, Jillian. 2018. "The Necropolitics of New World Nativism." *Early American Literature* 53, no. 3: 713–44.

Sedgwick, Catharine Maria. (1827) 1987. *Hope Leslie*. Edited by Mary Kelley. New Brunswick, NJ: Rutgers Univ. Press.

Seelye, John. 1985. "'Rational Exultation': The Erie Canal Celebration." *Proceedings of the American Antiquarian Society* 94, no. 2: 241–67.

Senior, Olive. 1977. "The Colon People." *Jamaica Journal* 11, nos. 3–4: 62–71.

Sheriff, Carol. 1996. *The Artificial River: The Erie Canal and the Paradox of Progress, 1817–1862*. New York: Hill and Wang.

Stowe, Harriet Beecher. (1841) 2003. "The Canal Boat." In *The Erie Canal Reader, 1790–1950*, edited by Roger W. Hecht, 96–102. Syracuse, NY: Syracuse Univ. Press.

Taylor, George Rogers. 1951. *The Transportation Revolution, 1815–1860*. New York: Rinehart.

Tocqueville, Alexis de. (1835) 2009. *Democracy in America*. Translated by Henry Reeve. Auckland, NZ: Floating Press.

Trump, Donald J. 2018. "President Trump Remarks on Infrastructure." *C-SPAN*, March 29. Video, 54:18. https://www.c-span.org/video/?443230-1/president-trump-delivers-remarks-infrastructure-richfield-ohio.

Wakefield, Stephanie. 2018. "Infrastructures of Liberal Life: From Modernity and Progress to Resilience and Ruins." *Geography Compass* 12, no. 7: 1–14.

Wertheimer, Eric. 1999. *Imagined Empires: Incas, Aztecs, and the New World of American Literature, 1771–1876*. New York: Cambridge Univ. Press.

Whyte, Kyle Powys. 2017a. "Indigenous Climate Change Studies: Indigenizing Futures, Decolonizing the Anthropocene." *English Language Notes* 55, nos. 1–2: 153–72.

Whyte, Kyle Powys. 2017b. "Our Ancestors' Dystopia Now: Indigenous Conservation and the Anthropocene." In *Routledge Companion to the Environmental Humanities*, edited by Ursula K. Heise, Jon Christensen, and Michelle Niemann, 206–15.

Whyte, Kyle Powys. 2018. "Settler Colonialism, Ecology, and Environmental Justice." *Environment and Society: Advances in Research* 9: 125–44.

Wolfe, Patrick. 2006. "Settler Colonialism and the Elimination of the Native." *Journal of Genocide Research* 8, no. 4: 387–409.

Wolff, Nathan. 2018. "The Weather in Dawson's Landing: Twain, Chesnutt, and the Climates of Racism." *American Literary History* 30, no. 2: 222–48.

Woollard, William E. 1937. Introduction to *New York—the Canal State* by Francis P. Kimball, vii–xii. Albany, NY: Argus Press.

Woodworth, Samuel. 1825. "Ode for the Canal Celebration." In Colden 1825: 250–53.

Wynter, Sylvia. 2003. "Unsettling the Coloniality of Being/Power/Truth/Freedom: Towards the Human, After Man, Its Overrepresentation—An Argument." *CR: The New Centennial Review* 3, no. 3: 257–337.

Jamin Creed Rowan The Hard-Boiled Anthropocene and the Infrastructure of Extractivism

Abstract This essay suggests that hard-boiled crime fiction in the United States has developed the kind of "deep infrastructural ethic" that John Durham Peters says is present in much modern thought. The essay attempts to illuminate the genre's infrastructural ethic and its corresponding affordance for environmental critique by tracing its expressions through a sample of significant texts in the hard-boiled and noir canons, and by concluding with a sustained reading of Paolo Bacigalupi's *The Water Knife* (2015). These readings demonstrate that hard-boiled narratives enable readers to perceive the ways in which extractivist infrastructures are frequently built upon and facilitate the exploitation of both human and environmental resources. Hard-boiled texts help readers see capitalism's extractivist infrastructure as a type of material and intellectual entrapment that ultimately undermines the common good and the planetary commons. Further, this essay argues that hard-boiled crime fiction attends to what AbdouMaliq Simone calls "infrastructures of relationality" and thus points a way out of the material and metaphysical entrapments of an extractivist economy's infrastructure. The infrastructures of relationality that emerge in a world in which climate crises have broken down the infrastructures of capitalism provide a platform from which individuals can practice a mode of collective thinking and being that offers an alternative to the alienation upon which extractivism depends. In short, the hard-boiled genre is not only one of the Anthropocene's earliest cultural responders but is also a vital genre for making sense of our contemporary situation in a deeper stage of the Anthropocene.

Keywords climate change, infrastructure studies, environmental humanities, genre studies

It would be easy to slot Paolo Bacigalupi's *The Water Knife* (2015) into the cli-fi canon—a body of texts that responds to a world transformed by climate change through the use of science fiction's narrative and formal strategies—and leave it at that.[1] But *The Water Knife* interweaves into its cli-fi DNA the conventions of another, more established genre: hard-boiled crime fiction. The novel's plot is driven by its protagonists' pursuit of a nineteenth-century document,

American Literature, Volume 93, Number 3, September 2021
DOI 10.1215/00029831-9361237   © 2021 by Duke University Press

hidden in a copy of Marc Reisner's *Cadillac Desert* (1986), that would entitle its possessors to "millions of acre-feet" (Bacigalupi 2015: 363) of Colorado River water—a lifeline at a moment when "Big Daddy Drought" has radically reconfigured the Southwest's social, political, and spatial infrastructures (22). The search for this recently discovered agreement among the Department of the Interior, the Bureau of Indian Affairs, and Pima tribal leaders drives the novel's two detective figures not only down Phoenix's "mean streets," to borrow Raymond Chandler's (1944: 59) shorthand for the genre he helped establish, but also in and out of the city's "six-lane boulevards" (Bacigalupi 2015: 28), "parking lots" (39), "subdivisions" (26), "condo complexes" (309), "disaster barrios" (257), "strip malls" (39), "water lines" (139), "water pumps" (236), and "gas pumps" (304). *The Water Knife* deploys the conventions of hard-boiled crime fiction to attend to the infrastructural "remains of a metropolitan sea" that once enveloped Phoenix—a "dust-draped sprawl of low-rises and abandoned single-families slumping across the flat desert basin" (68). In using hard-boiled crime fiction to investigate the infrastructural origins and consequences of climate change, Bacigalupi invites us to consider the utility of this genre at this particular moment in the Anthropocene.

While it is true, as Amitav Ghosh (2016: 7) notes, that the contemporary writer must face the challenges posed by climate change through the "grid of literary forms and conventions that came to shape the narrative imagination in precisely that period when the accumulation of carbon in the atmosphere was rewriting the destiny of the earth," it is not necessarily the case, contrary to Ghosh's suggestions, that those literary forms and conventions have been irretrievably "drawn into the modes of concealment that prevented people from recognizing the realities of their plight" (11). Hard-boiled crime fiction has, no doubt, done its fair share of concealing from its readers the infrastructures responsible for our planet's catastrophic accumulation of carbon. But its forms and conventions have, from their consolidation in the early twentieth century, also provided a grid upon which its practitioners have worked to expose the environmental consequences of what Timothy Mitchell (2011: 5) describes as our "carbon democracy" and what Naomi Klein (2014: 169) refers to as our "extractivist economy." That is to say, hard-boiled crime fiction provides us with what Stephanie LeMenager (2017: 222) calls the "patterns of expectation" and "means of living" that are capable of helping us more carefully address our current planetary emergencies. Bacigalupi's (2015: 140) use of hard-boiled crime fiction to address a world in which, as he puts it, "climate change and dust storms and fires and droughts" have flushed "#PhoenixDowntheTubes" (25) puts us in the position of recognizing the hard-boiled genre not only as one of the Anthropocene's earliest cultural responders but also as a vital genre for making sense of our situation in a much deeper stage of the Anthropocene than the one in which the genre emerged.

Scholars of hard-boiled crime fiction and the closely related film noir tradition have, of course, been talking about both genres' compelling critiques of capitalism and liberalism for quite some time, but these literary and cinematic genres have also been particularly good at exposing the environmental costs of US capitalism's extractivist infrastructure.[2] If it is true, as Patricia Yaeger (2007: 16) points out, that infrastructure is "present but barely visible" in much of our literature, then hard-boiled crime fiction and film noir are unusually attentive to the networks through which people, objects, information, and energy move and connect. Their detective protagonists pursue people, objects, information, and energy in and through subway trains, automobiles, railroads, highways, alleys, elevators, telephones, dictaphones, tract homes, canals, water pipelines, riverbeds, and oil fields. Hard-boiled novelists and noir directors might be said to practice what Geoffrey C. Bowker and Susan Leigh Star (1999: 34) call "infrastructural inversion"—the "struggle against the tendency of infrastructure to disappear"—or what Lisa Parks (2012: 65) refers to as "infrastructural intelligibility." I contend that hard-boiled crime fiction and film noir possess the kind of "deep infrastructural ethic," a commitment to making "environments visible," that John Durham Peters (2015: 35, 38) says is present in much modern thought.

In making the concealed infrastructure of capitalism more visible, hard-boiled and noir narratives expose the ways in which extractivist infrastructures are frequently built upon and facilitate the exploitation of environmental resources. The genres perform the type of cultural work that Jason W. Moore (2017: 606) and others advocate through their use of the term "Capitalocene" in the way that they identify "capitalism as a system of power, profit, and re/production in the web of life." As hard-boiled detectives navigate the city's infrastructure, they push back against the kind of infrastructural amnesia upon which those with power depend to acquire control and to cover up the environmental, social, and economic crimes that they commit to maintain their power. As Bacigalupi's use of the hard-boiled genre in *The Water Knife* demonstrates, the conventions that hard-boiled writers and noir directors have used to call attention to twentieth-century capitalism's effects upon individuals and communities might continue to be useful in interrogating acts of environmental exploitation. Given the ways in which scientists, politicians, activists, and public intellectuals have increasingly articulated their concerns about our current planetary emergencies through the rhetoric of criminality, it makes sense that crime narratives of various stripes would be among those to which culture makers would turn in their efforts to raise awareness about our environmental transgressions. It seems odd, then, that hard-boiled crime fiction's value in helping readers think through the types of complex crimes that emerged within what Sean McCann (2000: 16) describes as the twentieth-century's "corporate concentration of economic power" and "rapid growth of the state" has not moved environmental humanists to more thoroughly examine the genre's cultural capacity for sleuthing out the Capitalocene's infrastructural crimes.

This essay identifies the hard-boiled crime narrative's "affordance," to borrow Caroline Levine's (2015a: 6) term, for environmental critique by first tracing its expressions through a sample of significant texts in the hard-boiled and noir canons. As I demonstrate in the essay's first section, twentieth-century hard-boiled narratives such as Chandler's *The Big Sleep* (1939), James M. Cain's *Mildred Pierce* (1941), Ross Macdonald's *Sleeping Beauty* (1973), and Robert Towne and Roman Polanski's *Chinatown* (1974) display the hard-boiled crime narrative's infrastructural intelligibility—its ability to expose the ways in which extractivist infrastructures compromise the integrity of the planet by concealing capitalism's environmental criminality and over-determining its future. These and other practitioners of the genre establish a narrative structure with a deep infrastructural ethic that, as I suggest in the second section of the essay, Bacigalupi inhabits to make sense of twenty-first-century planetary emergencies. He uses the hard-boiled crime narrative's conventions, which had been developed by figures such as Chandler, Cain, Macdonald, Towne, and Polanski, to invite readers to see capitalism's extractivist infrastructure as a type of material and intellectual entrapment that ultimately undermines the common good and the planetary commons. I conclude the essay by arguing that Bacigalupi augments hard-boiled crime fiction's less pronounced commitment to attending to what AbdouMaliq Simone (2015: 18) calls "infrastructures of relationality" to point a way out of the material and metaphysical entrapments of an extractivist economy's infrastructure. The infrastructures of relationality that emerge in a world in which climate crises have broken down the infrastructures of capitalism provide a platform from which individuals can practice a mode of collective thinking and being that offers an alternative to the alienation upon which extractivism depends. In its use of hard-boiled crime fiction to both diagnose the crimes at the heart of the Capitalocene and to imagine alternative futures, Bacigalupi's *The Water Knife* persuasively demonstrates the facility of this genre's affordances at this particular moment of climate crisis.

The Infrastructure of the Hard-Boiled Crime Narrative

Although hard-boiled crime fiction's gritty realism is typically praised for exposing the deceptions propping up the nation's capitalist and democratic myths of progress, the genre has also been committed from its beginnings to calling attention to the infrastructures that perpetuate those myths. The gritty realism that Chandler, Cain, Dashiell Hammett, and others codified does the type of work that Levine (2015b) attributes to more mainstream realism—it unsettles the "privileged obliviousness that prevents some readers from noticing the crucial work of both structures and infrastructures in everyday life" (588) by describing "as sharply as possible the specific kind of order that each structure and infrastructure imposes on everyday life" (591). Through its descriptions of the mundane, Levine notes, realism has the potential to "estrange our routine of ignoring routine" (595). Hard-boiled crime fiction's brand of realism is particularly meticulous in its description of the structures and infrastructures of everyday life in a capitalist economy. Cain's *Mildred Pierce*—a novel that may not be a straightforward expression of the genre in that its narrative tails a mother whose "only crime, if she had committed one, was that she had loved" her daughter "too well" (Cain 1989: 297)—showcases the broader genre's deep commitments to describing the infrastructures that impose order on ordinary life in the mid-twentieth-century United States. Readers first encounter Bert Pierce mowing the lawn of his Spanish bungalow, which is a "lawn like thousands of others in southern California" (3). After mowing, Bert "[gets] out a coil of hose, screw[s] it to a spigot, and proceed[s] to water" his trees, grass, and tiled walk (4). Finished with his yardwork, Bert steps inside his home into a living room that "correspond[s] to the lawn he left"—it is the "standard living room sent out by department stores as suitable for a Spanish bungalow"—and then makes his way to the bathroom (4). "Twenty seconds" after he "tweak[s] the spigots" of the bathtub, he "step[s] into a bath of exactly the temperature he wanted, wash[es] himself clean, tweak[s] the drain, step[s] out, drie[s] himself on a clean towel, and step[s] into the bedroom again, without once missing a bar of the tune he [is] whistling, or thinking there [is] anything remarkable about it" (5). Cain's detailed account of Bert's suburban, domestic routine calls the reader's attention to the infrastructures upon which that routine rests: water pipelines, electrical grids, the lumber industry, furniture factories, and the railroad and street networks through which commodities make their way to consumers like Bert. Even before the narrative action picks up, Cain's narrator informs readers that Bert's infrastructural obliviousness constitutes the deeper crime within which the novel's surface crimes of blackmail, embezzlement, and rape will be situated. Bert's crime is his investment in a society that has "forgotten more than all other civilizations ever knew, in the realm of practicality" (5). Cain makes it clear that hard-boiled realism is committed to calling attention not just to the presence of infrastructure in the lives of its characters but also to the crime of forgetting the insidious social, political, economic, and environmental orders that capitalism's extractivist infrastructures impose upon its participants. In this section, I turn my attention to the way that hard-boiled crime fiction's infrastructural ethic exposes the physical and intellectual structures of extractivism's environmental order.

One of the most astute expressions of the genre's infrastructural intelligibility, Chandler's *The Big Sleep* begins and ends in an oil field. As Philip Marlowe leaves the Sternwood estate after his first meeting with General Sternwood, who hires him to look into a blackmail scheme involving his daughter Carmen, Chandler's iconic detective pauses on the front steps and looks out over the miles of sloping hills upon which the Sternwood home sits. Marlowe's eyes sweep over a "lower level faint and far off" where he can "just barely see some of the old wooden derricks of the oil field from which the Sternwoods had made their money" (Chandler 1992: 21). Although some of the wells are still pumping "five or six barrels a day," Marlowe observes that most of the oil field is "public park now, cleaned up and donated to the city by General Sternwood" (21). Standing on the Sternwood estate, Marlowe notes that the family can "no longer smell the stale sump water of the oil, but they could still look out of their front windows and see what had made them rich" if they "wanted to," though he doesn't "suppose they would" (21). After pursuing the blackmail scheme, which opens up onto an extensive network of crime and power, Marlowe finds himself, at the end of the novel, in the Sternwoods' oil field with Carmen. Witnessing the oil wells up close confirms Marlowe's initial observations: the place is "pretty creepy, all rusted metal and old wood and silent wells and greasy scummy sumps." Being there is, Marlowe reflects, "kind of eerie" (222). The oil field only becomes more eerie when Carmen attempts to shoot him—an act that is, Marlowe realizes, an echo of Carmen's earlier and successful attempt to shoot Rusty Regan, a "big curly-headed Irishman" (10) who had married her sister Vivian and who was the "breath of life" to General Sternwood (11). When she learns about Rusty's death, Vivian hires the racketeer Eddie Mars to deposit his body in the "sump" of the oil well, where it "decay[s]" as it bathes in the oil, chemicals, and other byproducts of oil extraction (229).

The Sternwoods' derricks and oil wells are just a small part of the much larger infrastructures of mobility and energy that radically reshaped Los Angeles over the first half of the twentieth century, but the accounts of the oil field that open and close *The Big Sleep* convey much about the nature of hard-boiled crime fiction's infrastructural intelligibility. The Sternwoods' slice of extractivist infrastructure smells for the reasons that Bruce Robbins (2007: 26) suggests: "infrastructure smells . . . because attention is not paid, because it is neglected." Infrastructure in *The Big Sleep* also smells because it is the site of the novel's most significant crimes. In imbricating Carmen's murder of Rusty with the Sternwoods' petroleum business, Chandler imbues the process of oil extraction with the kind of petulant violence that motivates Carmen to shoot Rusty after he refuses her sexual advances. According to the novel's logic, the oil industry treats the natural environment in much the same way that Carmen treats Rusty—as a resource whose reason for being is determined in relation to the personal profit that it might provide and whose being is therefore inherently disposable. The Sternwoods' oil field is the site at which their alienation—their insistence on "stand[ing] alone," to borrow from Anna Lowenhaupt Tsing (2015: 5), "as if the entanglements of living did not matter"—is expressed most pointedly. Although it is easy to lose in the shuffle of dead bodies, *The Big Sleep* delivers a rather straightforward environmental critique of the oil industry. The derricks and oil wells smell not just because they have been neglected but also because they are the site at which the Sternwoods have exploited the natural and human elements wrapped up in their economic entanglements.

Infrastructure simultaneously functions as a type of weapon through which individuals commit crimes in *The Big Sleep* and as a mechanism for concealing criminality. Infrastructure is, of course, prone to keeping a low profile. This "infrastructural concealment," Parks (2012: 66) explains, often works to "reinforce hierarchies of knowledge." But those who have accumulated wealth and power also use infrastructure to distance themselves from and cover up the misconduct through which they accumulate and maintain wealth and power. The public park that General Sternwood constructs on his oil field and then donates to the city literalizes a much broader effort to obscure his transgressions from both himself and the public. He hides the eerie infrastructure of his oil business behind infrastructure that ostensibly serves the common good. But the public park is not really an expression of penance; General Sternwood continues to profit from the oil that his wells produce each day. Rather, the public park allows the Sternwoods as well as the residents of Los Angeles to evade any misgivings about the extractivist economy upon which their personal and civic wealth has been built. As infrastructure that facilitates emotional and ethical evasion, the public park performs the same job carried out by much of the city's infrastructure. The expansive network of paved streets, the city's incandescent lights, the booming harbor, and the sea of bungalows that are built upon the backs of derricks and underground pipelines not only physically mask the oil industry's extractive infrastructure but also lull those who are, in LeMenager's (2014: 6) words, "living within oil" to overlook the environmental and human costs of extricating petroleum from the earth.[3] Los Angeles may exist, as Kazys Varnelis (2008: 9) notes, "by grace of infrastructure," but *The Big Sleep* makes it clear that this infrastructure graces residents, like General Sternwood, with the ability to avoid its smell and to circumvent an ethical reckoning with the processes that "made them rich" (Chandler 1992: 21).

While the infrastructural ethic on display in the first wave of hardboiled texts is capable of generating an environmental critique of capitalism's extractivist infrastructure, as it does in *The Big Sleep*, those who inherited the genre in the second half of the twentieth century more deliberately directed its affordance for infrastructural intelligibility to advance an environmental paradigm that emerged in the US environmental movement of the 1960s and 1970s. Hard-boiled writers such as Macdonald (1972: ix) used the genre to examine the nature of what he describes as an "ecological crime—a crime without criminals but with many victims"—and to raise questions about "whether we really have to go on polluting the sea and land and air in order to support our freeway philosophy of one man, one car." Through Lew Archer, the detective figure of many of his hard-boiled novels, Macdonald adapts the genre's conventions to investigate crimes that are much more explicitly framed as ecological. He transforms the detective figure into what Archer self-describes as a "bit of an environmentalist" (Macdonald 1973: 49). In *Sleeping Beauty*, a novel that opens with the image of an "offshore oil platform" that stands "up out of its windward end like the metal handle of a dagger that had stabbed the world and made it spill black blood" (1), Archer attempts to help readers think through the problem of being "hooked on oil" and of figuring out "how we can get unhooked before we drown in the stuff" while solving the more conventional crimes of murder, extortion, and kidnapping (110).[4] But Towne and Polanski's neo-noir film, *Chinatown*, pushes hard-boiled realism's propensities to expose the environmental costs of capitalism's extractivist economy about as far as anyone had up to that point. Towne (quoted in Erie 2006: 29), the film's screenwriter, explained that he used the conventions of noir to address the "great crimes in California" that had been "committed against the land—and against the people who own it and future generations." *Chinatown*, *Sleeping Beauty*, and other hard-boiled texts from the period showcase the hard-boiled crime narrative's evolution toward a sharper and more deliberate environmentalism—an environmentalism that remains particularly attentive to infrastructure and that is, as a result, poised to interrogate the crimes at the heart of the Capitalocene.

Chinatown is a particularly revealing case study of the environmental ends to which the genre's affordance for attending to infrastructure can be directed because it places infrastructure at the very center of its story about the great crimes that had been and continued to be committed against the land—an unusually acute infrastructural and environmental ethic that can be attributed not just to its hard-boiled plot but also to its cinematic medium. Along with other neo-noir films from the period, such as *Soylent Green* (1973) directed by Richard Fleischer and *Blade Runner* (1982) directed by Ridley Scott, *Chinatown*'s cinematic capacity for infrastructural inversion—for visually dragging infrastructure out into the open—unlocked new possibilities within the hard-boiled crime narrative for environmental critique. *Chinatown*'s central aesthetic project is to make Los Angeles's water infrastructure as visible to viewers as possible. The cinematography of John A. Alonzo takes viewers to a dry-as-a-bone section of the Los Angeles River, Echo Park Lake, the outfall at Point Fermin Park, the Oak Pass Reservoir and its runoff channel, the reflecting pool at the Mulwray residence, and irrigation ditches in the San Fernando Valley. The camera directs the viewer's gaze not just to the typical noir sites of femme fatales and dead bodies but also to

Los Angeles's extensive water infrastructure. By the end of the film, viewers might find themselves echoing a modified version of detective Jake Gittes's exasperation: "Jesus, water [infrastructure] again!" (Towne 1997: 18).

Chinatown explicitly fuses the crimes at the heart of its hard-boiled plot to the water infrastructure that it has made hypervisible to viewers. An early scene in the film lays out the core infrastructural dilemma that Los Angeles faces—a dilemma that generates the crimes around which the plot revolves. Addressing the city's councilmen, former mayor Sam Bagby explains what they all already know: "Remember— we live next door to the ocean, but we also live on the edge of the desert. Los Angeles is a desert community. Beneath this building, beneath every street, there's a desert. Without water the dust will rise up and cover us as though we'd never existed!" (10). The solution to the problem of expanding a city located in a desert is, for these Los Angelenos, to build a dam in the city's hinterland and to pipe the water to the city. In its efforts to convince citizens to fund the Alto Vallejo Dam and the Los Angeles Aqueduct as an "INVESTMENT IN THE FUTURE" (15), a corrupt coterie of municipal leaders and wealthy citizens begins dumping thousands of gallons of water out of the city's reservoirs every night; if the public believes that the city is in a drought, their thinking goes, residents are more likely to support the initiative to build the infrastructure necessary to "bring the water to L.A." (140). While these men dupe the public into believing that the city is in a drought, Noah Cross begins buying up property in the San Fernando Valley at depreciated prices, knowing that the city will incorporate the northwest valley and route much of the water from the new dam to the yet-to-be-built community in this valley. As viewers slowly pick up on the details of this complex scheme to appropriate the water supply from Los Angeles's hinterland, Towne and Polanski overlay this infrastructural crime plot with a more traditional crime narrative, imbuing the former with the moral meaning of the latter. Viewers learn late in the film that, after his wife had died and the Van der Lip Dam gave way, killing over five hundred individuals, Cross had a "breakdown" and raped his daughter, Evelyn (129). After running away to Mexico to escape her father, Evelyn had returned to Los Angeles and married her father's colleague, Hollis Mulwray, who took care of her and of the daughter, Katherine, born out of Cross's incestuous act. Given Evelyn's marriage to Mulwray—the chief engineer at Los Angeles's Department of Water and Power, who is murdered for trying to thwart the plan to build the Alto Vallejo Dam—*Chinatown* encourages viewers to see Cross's machinations to build water infrastructure as existing along the same ethical spectrum as his incestuous rape of Evelyn.

By imbricating Cross's felonious pursuit to construct water infrastructure with his rape of Evelyn, *Chinatown* delivers a brand of environmental critique tied to a long cultural tradition of feminizing the land and the corresponding rhetoric of despoliation. *Chinatown*'s environmentalist rhetoric has its roots in the historical accounts of the events upon which the film is, in part, based. After Los Angeles completed the construction of an aqueduct in 1913 to carry water over two hundred thirty miles from the Owens Valley to the San Fernando Valley, reporters such as Morrow Mayo (1932: 220) referred to this act as the "Rape of the Owens Valley." Mayo and others primarily used the term *rape* to signify the social cruelty of the city's chicanery. Los Angeles, according to Mayo, "destroyed a helpless agricultural section and a dozen towns" through its "ruthless, stupid, cruel, and crooked" behavior; it "deliberately ruined Owens Valley" by stealing its water supply and driving inhabitants "from their home" (246). The violence implied in Mayo's use of the phrase "The Rape of Owens Valley" has more to do with Los Angeles's unethical use of its power to "ruin" a "helpless" community than with its ecological transgressions. Because *Chinatown* closely aligns Evelyn and Katherine with the natural environment whose resources Los Angeles exploits to sustain life in an expanding desert metropolis, the film's closing scene, in which Lieutenant Escobar shoots Evelyn and Cross kidnaps Katherine in the streets of Los Angeles's Chinatown, signifies to viewers the human suffering and environmental violence that are the byproducts of infrastructural progress. *Chinatown* projects the violence inflicted upon its female protagonists onto the violence that Cross and others exact upon the Southern California landscape. By intertwining the rape plot with the infrastructure plot, *Chinatown* condemns the municipal leaders and powerful citizens who were behind the scheme to build a dam and aqueduct for appropriating a natural resource that belongs to another community without its consent. Of course, by gendering the environment against which these crimes are committed as distinctly female, *Chinatown* participates in a rhetorical tradition that, as Annette Kolodny would point out shortly after the film's release, invites the very kind of violence against the land that the film critiques; representing the "land as essentially feminine," Kolodny (1975: 4) notes, invites men to perceive possession of the land as the ultimate act of "gratification." It is, sadly, one of cinema's most tragic ironies that Polanski directed a film that would anticipate and condemn his own Crossesque rape of thirteen-year-old Samantha Jane Gailey in 1977—Polanski's own attempt at personal gratification through possession of the feminine.

The gratification that Cross seeks to extract from raping both his daughter and the landscape resides, to a large degree, upon his desire and ability to impose his vision of the future upon others—to overdetermine the futures of both his kin and his community through nonconsensual assertions of his power. This desire to own the future is what compels Cross to build an aqueduct and reservoir that will bring water to the San Fernando Valley. When Gittes queries Cross about his perplexing motivation to build this infrastructure ("Why are you doing it? How much better can you eat? What can you buy that you can't already afford?"), Cross responds matter-of-factly, "The future, Mr. Gittes—the future" (Towne 1997: 141). *Chinatown* sees infrastructure, like rape, as an attempt to overdetermine the future of others. The film reveals that infrastructures are always, as Hannah Appel, Nikhil Anand, and Akhil Gupta (2018: 27) suggest, "promises made in the present about our future." As materialized promises about the future, infrastructures evoke what Kregg Hetherington (2016: 42) describes as the "future perfect" tense in the way that they "suspend" the "present as the future's necessary past"; the "unfolding anticipation" that infrastructures conjure for their users, Hetherington explains, may function as an "aspirational mode" for some but lays a "temporal trap" for others "who are condemned to disappearance in an emerging order." The emerging suburban order that constitutes the future perfect of Cross's water infrastructure projects would preserve not only his individual power but also the type of white patriarchy that he seeks to solidify through rape. The suburban landscape that Cross hoped to bring into being through dams, reservoirs, and aqueducts would, he hoped, trap Evelyn, Katherine, and other females in increasingly tighter and controlled spaces. The suburban landscape that Cross's water infrastructure anticipates would also, with the help of redlining practices and racial covenants, keep people of color, such as Los Angeles's Asian inhabitants with whom the film aligns Evelyn and Katherine, trapped in the city's marginalized neighborhoods.[5] Determining the shape of the future's built environment through the construction of infrastructure enables Cross and his successors to maintain the structures of power from which they derive their authority by deciding who will and will not have access to the economic value extorted from the natural resources that his infrastructures extract. Cross's infrastructures lay the foundation of what Ashley Dawson

(2017: 231) characterizes as the planet's emerging regime of "climate apartheid," in which "those who are least responsible for climate chaos are made to pay the most dearly"—a demographic that tends to be drawn around racial, gender, economic, and geographic lines.

Chinatown invites viewers to think of the water infrastructure that Cross and his cronies con the public into constructing as a sequestering of natural resources from the distant future. Building dams, aqueducts, and reservoirs is, contrary to the promotional material about the bond distributed to citizens, not at all an investment in the future. Rather, these infrastructures introduce what Mitchell (2020) characterizes as a temporal "delay" that places the future "further away," thereby enabling extractivist capitalists like Cross to "capture revenue from the future and sell it at a discount in the present." As financial instruments that extract monetary value from natural resources at a fraction of their ecological value, Cross's water infrastructures subject future Los Angelenos to unpredictable environmental consequences and to the social, economic, and political byproducts of a destabilized climate. Capitalism's extractive infrastructures fail to consider the consent of those who will occupy the distant future, and the vitality of the environments that these future residents will inhabit, during the decision-making process of the present. *Chinatown* argues that infrastructure covers up future generations.

Climate Change and the Hard-Boiled Crime Narrative

When writers and culture makers began to more self-consciously and frequently address climate change through their art in the late twentieth and early twenty-first centuries, they looked to the hard-boiled crime and noir narratives of the 1930s and later for examples of how the genres might be deployed to address global warming and other planetary emergencies. Crime narratives are hardly perfect vessels for exploring the complex crimes that have propelled the planet from the Holocene to the Anthropocene. Crime narratives risk, as Sarah Dimick (2018: 20) notes, "exchanging the complicated spectrum of common yet differentiated responsibility for an easy division between guilt and innocence"; the concepts of "victimhood and villainy" that are at the heart of so many crime narratives are not, Dimick contends, "equipped to encapsulate the gradations and contradictions" of culpability for global warming that "exist across a region or country—or even within an individual human being" (30). But not all crime narratives make such a sharp distinction between victim and villain; hard-boiled crime fiction and film noir often divvy up culpability among a wide range of characters, including the detective figure. And the cultural work that hard-boiled crime fiction and film noir perform goes far beyond the pursuit of answers to whodunit questions. Bacigalupi's *The Water Knife* summons what *The Big Sleep* and *Chinatown* demonstrate—that the hard-boiled crime narrative is particularly useful for those seeking to address environmental crises through their art. The novel adeptly manipulates the genre's conventions to depict and diagnose the actualization of the future perfect that General Sternwood, Noah Cross, and others anticipated through the development of extensive networks of extractivist infrastructure.

The crimes that pepper Bacigalupi's hard-boiled novel are explicitly linked to climate change and the extractivist infrastructures that brought it into being. At the center of *The Water Knife*'s plot is the search for a nineteenth-century document that granted the Pima tribe access to "millions of acre-feet" of Colorado River water. Those searching for this document believe that the state that possesses these elusive water rights would be able to stave off the most dramatic consequences of global warming and would, consequently, become "arbiters of their own fate instead of a place of loss and collapse" (Bacigalupi 2015: 363). With "rights like these," Arizona's politicians and business elites believe that they could "rechannel the Colorado away from California, away from Nevada," and "pour water into a different set of deserts and a different set of cities" (363). The arrival of "Big Daddy Drought" (22) to the region creates the conditions in which the possession of water rights supersedes the protection of human rights—a world in which crime flourishes. The search for the Pima water rights generates a significant body count, including the torture and murder of a Phoenix water lawyer, whose death is at the center of the novel's plot. But the broader social conditions within which the central plot unfolds are also rife with violence set in motion by climate change. Lucy Monroe, a journalist who functions as one of the novel's two detective figures, makes a career out of documenting the dead bodies that turn up all over Phoenix—from the "swimmers" (corpses abandoned in empty swimming pools across the city) to the scores of Texan refugees buried in the desert by "predatory human traffickers" (65, 110). It doesn't take Lucy long to learn that twentieth-century infrastructure's "biopolitical promises" were, to use Dominic Boyer's (2018: 226) terms, "deluded and bankrupt, designs for Malthusian tragedy." While climate change typically generates what Rob Nixon (2011: 2) describes as a hard-to-represent "slow violence," *The Water Knife* suggests that the failure of infrastructure to fulfill its biopolitical promises can also manufacture the type of explosive violence that the hard-boiled detective was designed to pursue—the kind of violence that, in Bacigalupi's (2015: 111) words, "got American bureau chiefs salivating and news teams on the next plane out."

While *The Water Knife* points a finger at Big Daddy Drought for the violence that those living in the Southwest experience as they move deeper into the Capitalocene, the novel is especially interested in directing blame for the crimes that it chronicles at what Bacigalupi calls "development boosterism" (103).[6] Cities built upon the ideology of "development boosterism" are, according to Bacigalupi, not simply backdrops upon which the devastations of climate change will play out. *The Water Knife* suggests instead that cities are, as Harriet Bulkeley (2013: 4) puts it, "central to the ways in which the vulnerabilities and risks of climate change are produced." In these particular climatic conditions, a Phoenix cop informs Lucy, the "whole city's a suspect" (Bacigalupi 2015: 65) of committing crimes against its inhabitants—of repeatedly putting them in vulnerable and risky positions—at least in part because industrial and postindustrial cities are environmentally unsustainable and are therefore always susceptible to "collapse" (72). The novel's repeated use of the term *collapse* to describe the effect of climate change upon cities such as Phoenix draws attention to the particular type of violence that the US city inflicts upon its residents. If cities have, over the last two centuries, entangled urbanites in a web of what Timothy Clark (2015: 159) calls "material entrapment" that forces them to rely upon infrastructures that assume unlimited access to natural resources (water, petroleum, coal, food), the diminishment of those resources in a period of intensifying global warming means that cities impose an even more constrictive form of entrapment upon their occupants. Bacigalupi imagines a Phoenix that collapses upon its residents by failing to provide them with viable alternatives to a lifestyle dependent upon the intensive consumption of fossil fuel and water. That is to say, the city collapses the realm of agency within which its inhabitants operate, constricting their ability to adapt to new environmental circumstances. Bacigalupi (2015: 26) also imagines a future in which Phoenix physically collapses, in which the city is a type of "sinkhole, sucking everything down—buildings, lives, streets, history—all of it tipping and spilling into the gaping maw of disaster—sand, slumped saguaros, subdivisions—all of it going down." *The Water Knife* frames the city's conventional signifiers of stability—its buildings, streets, subdivisions, and other infrastructures—as unstable to the point that the very structures built to shelter residents ultimately collapse upon and entrap them.

In depicting the city as the perpetrator of criminal activity, *The Water Knife* does what hard-boiled and noir narratives have been doing for nearly a century—it challenges readers to see the urban infrastructure of capitalism as the undoing of the public's wellbeing. The novel does not produce, to borrow Michael Rubenstein's (2010: 7) terminology, a "weak-messianic story about the development of public utilities" that depicts infrastructure as the foundation of the "common good." Contrary to Rubenstein, Bacigalupi suggests that the vision of the common good at the center of twentieth-century urban development in the western United States rests upon networks of infrastructure that assume an infinite supply of water, oil, and other natural resources. But when the "water [runs] out," as it has in the world of *The Water Knife*, and residents of cities and towns throughout the West realize "too late that their prosperity was borrowed," the infrastructure that had propped up their prosperity becomes not only unusable but also menacing (Bacigalupi 2015: 20). Every town that had depended upon the Colorado River for its water looks the same in the novel's postwater landscape: "traffic lights swinging blind on tumbleweed streets; shadowy echoing shopping malls with shattered window displays; golf courses drifted with sand and spiked with dead stick trees" (51). The spaces designed to facilitate mobility, consumption, and leisure now taunt residents with their uselessness; the sharp edges of the shopping mall's "shattered" glass and the golf course's "dead stick" trees express the threatening resistance of an extractivist economy's infrastructure to being repurposed for alternative versions of the common good. The spatial and economic value systems supported by the infrastructure of capitalism's environmental calculus inevitably land the public in social, financial, and ecological insolvency. All those who had "owned" and "felt rich living in their 5 Bed/3 Bath houses" in Phoenix's suburban subdivisions are, when Phoenix "shut[s] off their water," left with "granite countertops" that are "now just polished rock no one [gives] a shit about." In an extractivist economy, the egregious but hidden environmental speculation upon which the "resale value" of any asset depends leaves those who participate in that economy exposed to irrecoverable impoverishment (225). The present tense of any moment in the Capitalocene has, Bacigalupi suggests, been ecologically looted by the infrastructure of the past and is, therefore, subject to a growing degree of social, economic, and environmental volatility.

The infrastructure of capitalism undermines the common good of those inhabiting the world of *The Water Knife* not just because it cannot be adapted to the climatic conditions to which it has contributed but also because it prevents those who depend upon it from perceiving what the common good is. Bacigalupi contends that infrastructure built to advance an extractivist economy blinds those who inhabit the spaces it creates to the shifting dimensions of the common good by concealing from them the inextricable relationship between the common good and the environmental commons. Infrastructure cultivates what Maria, one of Phoenix's many Texan refugees, refers to as "*ojos viejos*"—old eyes that keep her father and others behaving "according to an ancient map of the world that no longer exist[s]" (42). The "map of the world" through which Maria's father and other Phoenix residents see their environment and those who inhabit it is shaped largely by the infrastructures that move goods, resources, people, and ideas through the larger "metropolitan sea" in which the city is situated (68). Infrastructure operates like language in the sense that possessing the "right words," according to Bacigalupi, determines our ability or inability to see "what is right in front of our faces" (59). The "poetics of infrastructure," to borrow Brian Larkin's (2013: 329) terminology, not only "address[es] and constitute[s] subjects" in the moment of its formulation by communicating a dominant culture's "desire and fantasy" but also perpetuates that paradigm far beyond the temporality of its origins. Among the many assumptions that an extractivist economy's infrastructure propagates is the "illusion" under which Lucy (and so many of her neighbors) "had been living"—that "she could keep herself separate" both from her fellow urbanites and from the environment upon which they all depend (Bacigalupi 2015: 290). As the conventional detective figure of Bacigalupi's hard-boiled narrative, Lucy may inherit her lone-wolf inclinations from her fictional predecessors, but the infrastructure spawned by Phoenix's development boosterism deepens her delusions of autonomy.[7] Like the "preppers," for whom the illusion of separateness has become so complete that they stockpile resources and plan to "ride out the apocalypse alone" in the desert (329), Lucy struggles to see beyond her consuming desire to "divine something meaningful from this place's suffering" (272) and to acknowledge that the Sonoran Desert was, in fact, a "stupid place to grow a city" (164–65). She cannot recognize that pursuing the common good might require abandoning Phoenix rather than trying to save it. The infrastructure of capitalism is menacing, then, not just because it fails to sustain the common good but also because it promotes a hard-to-shake blindness to what the common good might look like in the age of climate change. Bacigalupi contends that the poetics of capitalism's extractivist infrastructure hardwires its assumptions about unlimited access to natural resources into its users to the degree that they seem incapable of reimagining the dimensions of the common good in an environmental context in which the availability of natural resources is clearly limited. This infrastructurally induced blindness is the deepest crime in *The Water Knife*'s hardboiled world.

Infrastructures of Hard-Boiled Relationality

The Water Knife, like many of its hard-boiled and noir predecessors, calls attention not only to the ways in which capitalism's extractive infrastructure exploits human and environmental resources but also to the "relational infrastructures," to borrow Simone's (2015: 18) terminology, that emerge within the interstices of the urban spaces and social formations engineered by the infrastructures of the Capitolocene. These "infrastructures of relationality," according to Simone, "construct circulations of bodies, resources, affect and information" that are capable of transgressing the circuits of the dominant spatial and social orders (18). Infrastructures of relationality serve as "tools through which political imaginations and claims are exerted" because they keep open the "many different trajectories of what life could be" (18). From the early twentieth century, writers and filmmakers have deployed the infrastructural ethic of hard-boiled and noir narratives to illuminate the infrastructures of relationality that give life to marginalized social assemblages and to their alternative "political imaginations." Writers such as Chandler, Cain, and Hammett redeveloped the nineteenth-century detective, as McCann (2000: 158) notes, to be a figure capable of absorbing the "injuries" inflicted upon the city's most vulnerable inhabitants by an increasingly powerful state and the consolidated corporations it subsidizes. The infrastructure of relationality forged by early hard-boiled detectives like Philip Marlowe, McCann proposes, "salvages the demotic and mundane features of everyday life" to bring about an "otherwise suppressed representation" (158). From their origins, then, hard-boiled crime fiction and noir narratives have been interested both in exposing the concealed consequences of capitalism's infrastructure and in bringing to light the infrastructures of relationality that might point a way out of the material and metaphysical entrapments forged by an extractivist economy's infrastructure.

As Bacigalupi demonstrates in *The Water Knife*, the varieties and possibilities of relational infrastructures proliferate in a world in which climate crises have broken down capitalism's extractivist infrastructures and created space for alternative infrastructures to emerge. At a time when many cities are "losing their balance as the ground they'd taken for bedrock shift[s] beneath them" (Bacigalupi 2015: 72), some in Bacigalupi's arid Southwest respond to this breakdown by developing new infrastructures that strive to perpetuate capitalism's basic social and economic orders—to reinforce the lines of climate apartheid. The arcologies that sprout up in Phoenix and Las Vegas to shelter privileged urbanites may recycle water, rely entirely upon solar energy, and grow food on site, but these infrastructural expressions of sustainability only exacerbate the racial and class divisions of the early Capitalocene; one still needs the right "swipe cards," "fingerprints," or "friends" to make it past the arcologies' "security guards" (89). The climate resilience on display in the arcologies excludes Phoenix's predominantly Latinx population through architectural efforts to control and foreclose the city's infrastructures of relationality. But the infrastructures that spring up in the "drought-savaged wilderness of the Phoenix suburbs" (39), by contrast, are assembled by "stripping subdivisions," "hauling the scrap closer" to wherever the Red Cross had installed water pumps, and "packing housing into the sprawl that Phoenix had left wide and open" (238). Unlike "development boosterism," a hyped-up version of which drives the arcology construction boom in downtown Phoenix and Las Vegas, the ethic of development at work in the proliferation of squatter camps grows out of a commitment to turning "scavenged" material into "makeshift," "hacked-together," "ad-hoc" infrastructure; it is an ethic of "repurposing" (241). This repurposed infrastructure does not necessarily make for a comfortable or a secure life but can easily be modified to accommodate a wide range of users and uses.

The high-density "plywood ghettos" that hug the water pumps in Phoenix's suburban sprawl provide an "oasis of life and activity" for the climate refugees who inhabit them because their makeshift infrastructure brings into being new relational infrastructures (39). These refugees practice what Steven J. Jackson (2014) refers to as "broken world thinking" through the tactics of "repair" (174)—the "subtle acts of care" through which the refugees maintain and transform "order and meaning in complex sociotechnical systems" and through which they preserve and extend "human value" (175). The distinct circulatory paths that bodies, resources, information, and feelings take within these remade suburban neighborhoods make it possible for at least some of the climate refugees to better navigate the dominant social and spatial orders of climate apartheid. The repurposed suburban neighborhood surrounding the Red Cross/China Friendship pump, for instance, permits climate refugees who cannot "afford to keep cars" to live in a place where they can easily "catch a bus" and where they can "get water without having to walk so far" (Bacigalupi 2015: 241). These suburbanites clog the empty parking lots and streets that had once serviced "McMansions and strip malls" with "prayer tents" in which they beg "for salvation" and post the "numbers and names and pictures of loved ones they had lost" during their treacherous migration to Phoenix (39). Residents of the squats forge alternative economic relationships with one another as they sell "PowerBars and black-market humanitarian rations" (239) as well as "burritos and *pupusas* and soft tacos" (257) to one another in the improvised markets that spring up in the water pump's informal plaza. Because this ad hoc infrastructure does not overdetermine the ways in which Bacigalupi's climate refugees use and inhabit the city, they do what Simone (2004: 410) suggests many urbanites in similar situations do—they turn "commodities, found objects, resources, and bodies into uses previously unimaginable" in their efforts to "derive maximal outcomes from a minimal set of elements" (411). Their urban improvisations induct them into what Tsing (2015) describes as "shifting assemblages" (20) and "open-ended gatherings" that "don't just gather lifeways" but "make them" (23). The infrastructures of relationality that these assemblages of climate refugees build together may be tenuous and may constantly shift, but they do provide a platform upon which individuals might cobble together cultural, economic, and political lifeways that respond in some way to "all the horrors the world [has] to offer" (Bacigalupi 2015: 272). These suburbanites do their best to repair the harm inflicted upon them by capitalism's extractivist infrastructures.

Although the maximal outcomes that Bacigalupi's climate refugees generate through the minimal elements that they scavenge in Phoenix's "disaster barrios" (257) may not overthrow the social and economic orders of climate apartheid, the infrastructures of relationality that emerge in these neighborhoods sustain a mode of collective being that fundamentally undermines the alienation that feeds an extractivist culture. When Lucy and Angel retreat to a squat near the Red Cross/China Friendship pump after their search for the Pima water rights riles up private interests invested in obtaining these rights, they find themselves embedded in a very different set of assemblages than those in which they had previously been enmeshed. They inhabit what Janet Vertesi (2014: 269) refers to as the "seams" that connect overlapping infrastructural orders and, in so doing, realize that there are "many possible ways to patch multiple systems together into local alignment." In their new neighborhood, Angel finds that he is no longer the "lone individual" that he had insisted on being when he worked as a hitman for the Southern Nevada Water Authority. He realizes that he is instead inextricably aligned with others, is always "becoming invisible" as he mixes with the climate refugees around him (Bacigalupi 2015: 239). As Lucy becomes more entangled in the neighborhood's assemblages, she realizes that her previous belief that she could "just cover this place" through her journalism without being affected was wrongheaded and acknowledges that she is "part of it" (324). As temporary residents of this very different neighborhood, both Lucy and Angel not only practice broken-world thinking but also what Dipesh Chakrabarty (2009: 213) refers to as "species thinking," a mode of relationality that situates individuals in a much more expansive notion of "human collectivity" (222).The seams into which Angel and Lucy step as they move through this repurposed Phoenix neighborhood put them in a position to occupy new relational infrastructures and to experience the truth that one of their new neighbors speaks: "We're all each other's people" (Bacigalupi 2015: 250). While neither Lucy nor Angel translates the species thinking that they briefly inhabit into a sustained political project that provides a viable alternative to Phoenix's regime of climate apartheid, they do fundamentally understand that, as Angel puts it, "nobody survives on their own" (329). The infrastructures of relationality that develop within the Red Cross/China Friendship pump neighborhood offer an alternative to the way of thinking about the world that the poetics of capitalism's extractivist infrastructures continue to propagate.

In its use of hard-boiled crime fiction to both diagnose the infrastructural crimes at the heart of the Capitalocene and to imagine alternative futures, Bacigalupi's *The Water Knife* persuasively demonstrates not only the value of this particular genre but also of genre fiction more generally to help us navigate our current climate crises. In addition to its affordance for infrastructural intelligibility, which writers have developed and sharpened over time, hard-boiled crime fiction, like other types of genre fiction, feels especially useful at this moment because of the ways in which its storytelling supra-structure—its recognizable conventions and patterns—lends itself to the ethic of repurposing and retrofitting that we and the climate refugees in *The Water Knife* must practice to survive. The flexibility of genre fiction's narrative supra-structures, which stands in stark contrast to the inflexibility of extracti-vist infrastructures, makes it possible for writers like Bacigalupi to play with and within genre in an attempt to reconstruct not only infrastruc-tures of relationality but also the physical infrastructures needed to sus-tain alternative assemblages of the human and nonhuman. The struc-tural plasticity of hard-boiled crime fiction and other types of genre fiction facilitates the work that writers and other culture makers must undertake to help us renovate the dreams that new infrastructures can promise to fulfill.

Jamin Creed Rowan is an associate professor of English at Brigham Young Univer-sity. He operates at the intersection of literary studies, urban studies, and the envi-ronmental humanities. He is the author of *The Sociable City: An American Intellectual Tradition* (2017) and is currently at work on his next book project, "The Anthropo-cene's Urban Imaginary."

Notes

I am grateful to the American Studies Faculty Research Group at Brigham Young University (especially Brian Roberts, Mary Eyring, George Handley, and Sam Jacob) and members of my graduate seminar on "The Culture of Cli-mate Change" for their insightful and generous responses to earlier drafts of this essay.

1   Dan Bloom, a journalist and environmental activist, coined the term *cli-fi* in the late 2000s. The term really took hold in the popular imagination in 2013 when NPR ran a five-minute segment on its Weekend Edition about Nathaniel Rich's *Odds Against Tomorrow* (2013) and used the term *cli-fi* to describe the novel (Evancie 2013). Cli-fi is broadly used to refer to works of fiction about the climate but, perhaps because of its resonance with *sci-fi* and the sheer number of science fiction texts that address cli-mate change, is often used to refer more specifically to works of science fiction about environmental issues. For more thorough accounts of the cli-fi genre, see Irr 2017, LeMenager 2017, and Streeby 2018.

2   Some of the earliest French critics of American film noir (hard-boiled fic-tion's cinematic counterpart) characterized the tradition, according to James Naremore (2002: viii), as a "critique of savage capitalism." Scholars of the genre working in the Marxist tradition, such as Mike Davis (1990: 21), claim that "*noir* everywhere insinuated contempt for a depraved busi-ness culture." Recent scholarship of the genre has been more precise in articulating the type or features of capitalism that hard-boiled fiction and

the noir tradition critique. Postwar American novels and films in these genres examine the consequences of what Andrew Pepper (2011: 91) describes as the "Fordist regime of capital accumulation," what Christopher Breu (2005: 5) calls "corporate capitalism," and what Erik Dussere (2014: 25) characterizes as "consumer capitalism." These and other scholars of the hard-boiled crime narrative and noir tradition have primarily been interested in the consequences of capitalism upon what Susanna Lee (2016: 2) terms the "moral authority" of individual citizens—specifically, according to Megan E. Abbott (2002: 2), the "solitary white man"—and upon what Leonard Cassuto (2009: 11) labels the "communities of affiliation and consent" to which these individuals belong.

3 For an account of the history of oil infrastructure in the Los Angeles Basin, see Ruchala 2008.
4 For a more thorough assessment of Macdonald's use of hard-boiled fiction to explore environmental issues, see Ashman 2018.
5 For a thorough history of the federal regulations, banking practices, and tactics of homeowner associations that kept US suburbs white during the mid-twentieth century, see Rothstein 2017.
6 Phoenix has, according to Andrew Ross (2011: 15), "channeled the national appetite for unrestrained growth" more than "any other U.S. metropolis in the postwar period."
7 Lee (2016: 3) explains that hard-boiled crime fiction's deep investment in the "moral authority" of the "autonomous self" is expressed most explicitly in the detective figure.

References

Abbott, Megan E. 2002. *The Street Was Mine: White Masculinity in Hardboiled Fiction and Film Noir*. New York: Palgrave Macmillan.
Appel, Hannah, Nikhil Anand, and Akhil Gupta. 2018. "Introduction: Temporality, Politics, and the Promise of Infrastructure." In *The Promise of Infrastructure*, edited by Nikhil Anand, Akhil Gupta, and Hannah Appel, 1–38. Durham, NC: Duke Univ. Press.
Ashman, Nathan. 2018. "Hard-Boiled Ecologies: Ross Macdonald's Environmental Crime Fiction." *Green Letters: Studies in Ecocriticism* 22, no. 1: 43–54.
Bacigalupi, Paolo. 2015. *The Water Knife*. New York: Vintage.
Bowker, Geoffrey C., and Susan Leigh Star. 1999. *Sorting Things Out: Classification and Its Consequences*. Cambridge, MA: MIT Press.
Boyer, Dominic. 2018. "Infrastructure, Potential Energy, Revolution." In *The Promise of Infrastructure*, edited by Nikhil Anand, Akhil Gupta, and Hannah Appel, 223–43. Durham, NC: Duke Univ. Press.
Breu, Christopher. 2005. *Hard-Boiled Masculinities*. Minneapolis: Univ. of Minnesota Press.
Bulkeley, Harriet. 2013. *Cities and Climate Change*. New York: Routledge.

Cain, James M. (1941) 1989. *Mildred Pierce*. New York: Random House.

Cassuto, Leonard. 2009. *Hard-Boiled Sentimentality: The Secret History of American Crime Stories*. New York: Columbia Univ. Press.

Chakrabarty, Dipesh. 2009. "The Climate of History: Four Theses." *Critical Inquiry* 35, no. 2: 197–222.

Chandler, Raymond. 1944. "The Simple Art of Murder." *Atlantic Monthly*, December, 53–59.

Chandler, Raymond. (1939) 1992. *The Big Sleep*. New York: Random House.

Clark, Timothy. 2015. *Ecocriticism on the Edge: The Anthropocene as a Threshold Concept*. London: Bloomsbury Academic.

Davis, Mike. 1990. *City of Quartz: Excavating the Future in Los Angeles*. New York: Verso.

Dawson, Ashley. 2017. *Extreme Cities: The Peril and Promise of Urban Life in the Age of Climate Change*. New York: Verso.

Dimick, Sarah. 2018. "From Suspect to Species: Climate Crime in Antti Tuomainen's *The Healer*." *Mosaic* 51, no. 3: 19–35.

Dussere, Erik. 2014. *America Is Elsewhere: The Noir Tradition in the Age of Consumer Culture*. New York: Oxford Univ. Press.

Erie, Steven P. 2006. *Beyond Chinatown: The Metropolitan Water District, Growth, and the Environment in Southern California*. Stanford, CA: Stanford Univ. Press.

Evancie, Angela. 2013. "So Hot Right Now: Has Climate Change Created A New Literary Genre?" *National Public Radio*, April 20. http://www.npr .org/2013/04/20/176713022/so-hot-right-now-has-climate-change-created -a-new-literary-genre.

Ghosh, Amitav. 2016. *The Great Derangement: Climate Change and the Unthinkable*. Chicago: Univ. of Chicago Press.

Hetherington, Kregg. 2016. "Surveying the Future Perfect: Anthropology, Development and the Promise of Infrastructure." In *Infrastructures and Social Complexity: A Companion*, edited by Penelope Harvey, Casper Bruun Jensen, and Atsuro Morita, 40–50. London: Routledge.

Irr, Caren. 2017. "Introduction to Climate Fiction in English." In *Oxford Research Encyclopedia of Literature*, edited by Paula Rabinowitz. Oxford: Oxford Univ. Press. http://literature.oxfordre.com/view/10.1093/acrefore/97801 90201098.001.0001/acrefore-9780190201098-e-4.

Jackson, Steven J. 2014. "Rethinking Repair." In *Media Technologies: Essays on Communication, Materiality, and Society*, edited by Tarleton Gillespie, Pablo J. Boczkowski, and Kirsten A. Foot, 174–89. Cambridge, MA: MIT Press.

Klein, Naomi. 2014. *This Changes Everything: Capitalism vs. the Climate*. New York: Simon and Schuster.

Kolodny, Annette. 1975. *The Lay of the Land: Metaphor as Experience and History in American Life and Letters*. Chapel Hill: Univ. of North Carolina Press.

Larkin, Brian. 2013. "The Politics and Poetics of Infrastructure." *The Annual Review of Anthropology* 42: 327–43.

Lee, Susanna. 2016. *Hard-Boiled Crime Fiction and the Decline of Moral Authority*. Columbus: Ohio State Univ. Press.

LeMenager, Stephanie. 2014. *Living Oil: Petroleum Culture in the American Century*. New York: Oxford Univ. Press.

LeMenager, Stephanie. 2017. "Climate Change and the Struggle for Genre." In *Anthropocene Reading: Literary History in Geologic Times*, edited by Tobias Menley and Jesse Oak Taylor, 220–38. University Park: Pennsylvania State Univ. Press.

Levine, Caroline. 2015a. *Forms: Whole, Rhythm, Hierarchy, Network*. Princeton, NJ: Princeton Univ. Press.

Levine, Caroline. 2015b. "'The Strange Familiar': Structure, Infrastructure, and Adichie's *Americanah*." *Modern Fiction Studies* 61, no. 4: 587–605.

Macdonald, Ross. 1972. Introduction to *Black Tide: The Santa Barbara Oil Spill and Its Consequences*, by Robert Easton, ix–xvi. New York: Delacorte.

Macdonald, Ross. 1973. *Sleeping Beauty*. New York: Alfred A. Knopf.

Mayo, Morrow. 1932. *Los Angeles*. New York: Alfred A. Knopf.

McCann, Sean. 2000. *Gumshoe America: Hard-Boiled Crime Fiction and the Rise and Fall of New Deal Liberalism*. Durham, NC: Duke Univ. Press.

Mitchell, Timothy. 2011. *Carbon Democracy: Political Power in the Age of Oil*. New York: Verso.

Mitchell, Timothy. 2020. "Infrastructures Work on Time." In *New Silk Roads*, edited by Aformal Academy and *e-flux Architecture*. https://www.e-flux.com/architecture/new-silk-roads/312596/infrastructures-work-on-time/.

Moore, Jason W. 2017. "The Capitalocene, Part I: On the Nature and Origins of Our Ecological Crisis." *Journal of Peasant Studies* 44, no. 3: 594–630.

Naremore, James. 2002. "A Season in Hell or the Snows of Yesteryear?" In *A Panorama of American Film Noir (1941–1953)*, edited by Raymond Borde and Etienne Chaumeton, translated by Paul Hammond, vii–xxi. San Francisco: City Lights Books.

Nixon, Rob. 2011. *Slow Violence and the Environmentalism of the Poor*. Cambridge, MA: Harvard Univ. Press.

Parks, Lisa. 2012. "Technostruggles and the Satellite Dish: A Populist Approach to Infrastructure." In *Cultural Technologies: The Shaping of Culture in Media and Society*, edited by Göran Bolin, 64–84. London: Routledge.

Pepper, Andrew. 2011. "Post-War American Noir: Confronting Fordism." In *Crime Culture: Figuring Criminality in Fiction and Film*, edited by Brian Nicol, Eugene McNulty, and Patricia Pulham, 90–106. New York: Continuum.

Peters, John Durham. 2015. *The Marvelous Clouds: Toward a Philosophy of Elemental Media*. Chicago: Univ. of Chicago Press.

Polanski, Roman, dir. 1974. *Chinatown*. Los Angeles: Paramount Pictures.

Robbins, Bruce. 2007. "The Smell of Infrastructure: Notes toward an Archive." *boundary 2* 34, no. 1: 25–33.

Ross, Andrew. 2011. *Bird on Fire: Lessons from the World's Least Sustainable City*. New York: Oxford Univ. Press.

Rothstein, Richard. 2017. *The Color of Law: A Forgotten History of How Our Government Segregated America.* New York: Liveright.

Rubenstein, Michael. 2010. *Public Works: Infrastructure, Irish Modernism, and the Postcolonial.* South Bend, IN: Univ. of Notre Dame Press.

Ruchala, Frank. 2008. "Crude City: Oil." In *The Infrastructural City: Networked Ecologies in Los Angeles,* edited by Kazys Varnelis, 54–64. New York: Actar.

Simone, AbdouMaliq. 2004. "People as Infrastructure: Intersecting Fragments in Johannesburg." *Public Culture* 16, no. 3: 407–29.

Simone, AbdouMaliq. 2015. "Relational Infrastructures in Postcolonial Urban Worlds." In *Infrastructural Lives: Urban Infrastructure in Context,* edited by Stephen Graham and Colin McFarlane, 17–38. London: Routledge.

Streeby, Shelley. 2018. *Imagining the Future of Climate Change: World-Making through Science Fiction and Activism.* Berkeley: Univ. of California Press.

Towne, Robert. 1997. *Chinatown.* New York: Faber and Faber.

Tsing, Anna Lowenhaupt. 2015. *The Mushroom at the End of the World: On the Possibility of Life in Capitalist Ruins.* Princeton, NJ: Princeton Univ. Press.

Varnelis, Kazys. 2008. "Networked Ecologies." In *The Infrastructural City: Networked Ecologies in Los Angeles,* edited by Kazys Varnelis, 6–16. New York: Actar.

Vertesi, Janet. 2014. "Seamful Spaces: Heterogeneous Infrastructures in Interaction." *Science, Technology, and Human Values* 39, no. 2: 264–84.

Yaeger, Patricia. 2007. "Dreaming of Infrastructure." *PMLA* 122, no. 1: 9–26.

Suzanne F.
Boswell

"Jack In, Young Pioneer":
Frontier Politics, Ecological Entrapment,
and the Architecture of Cyberspace

Abstract This essay uncovers the environmental and historical conditions that played a role in cyberspace's popularity in the 1980s and 1990s. Tracing both fictional and critical constructions of cyberspace in a roughly twenty-year period from the publication of William Gibson's Sprawl trilogy (1984–1988) to the Telecommunications Act of 1996, this essay argues that cyberspace's infinite, virtual territory provided a solution to the apparent ecological crisis of the 1980s: the fear that the United States was running out of physical room to expand due to overdevelopment. By discursively transforming the technology of cyberspace into an "electronic frontier," technologists, lobbyists, and journalists turned cyberspace into a solution for the apparent American crisis of overdevelopment and resource loss. In a period when Americans felt detached from their own environment, cyberspace became a new frontier for exploration and a so-called American space to which the white user belonged as an indigenous inhabitant. Even Gibson's critique of the sovereign cyberspace user in the Sprawl trilogy masks the violence of cybercolonialism by privileging the white American user. Sprawl portrays the impossibility of escaping overdevelopment through cyberspace, but it routes this impossibility through the specter of racial contamination by Caribbean hackers and Haitian gods. This racialized frontier imaginary shaped the form of internet technologies throughout the 1990s, influencing the modern user's experience of the internet as a private space under their sovereign control. In turn, the individualism of the internet experience restricts our ability to create collective responses to the climate crisis, encouraging internet users to see themselves as disassociated from conditions of environmental and social catastrophe.

Keywords internet, science fiction, William Gibson, ecocriticism

When the internet opened to the public in the late 1980s, critics and technologists alike exalted the digital realm as a means of returning to a pristine nature. N. Katherine Hayles argued that "in a world despoiled by overdevelopment, overpopulation and time-release environmental poisons . . . a cyberspace body, like a cyberspace landscape, is immune to blight and corruption" (1999: 36).

American Literature, Volume 93, Number 3, September 2021
DOI 10.1215/00029831-9361251   © 2021 by Duke University Press

Scholars saw cyberspace as more than an artificial environment that was reminiscent of a state of nature because of its freedom from pollution.[1] In one of the earliest studies on the new field of cyberspace, professor of architecture Michael Benedikt (1991: 3) claimed the digital world had the potential to "decontaminat[e] the natural and urban landscapes . . . saving them from the chain-dragging bull-dozers of the paper industry, from the diesel smoke of courier and post office trucks, from jet fuel fumes and clogged airports . . . from all the inefficiencies, pollutions (chemical and informational), and corruptions attendant to the process of moving information attached to things." Like a virtual escape valve that opens at just the right time, cyberspace's very presence seemed to decontaminate the physical world, decreasing overpopulation, overdevelopment, and pollution by moving human excess into a digital environment.

Over thirty years later, contemporary ecocritics have revealed cyberspace's deep and continued connection to our physical world. Tracing the impact of telecommunication services, data centers, electricity, and cables, scholars like Tung-Hui Hu, Allison Carruth, and Nicole Starosielski challenge the internet's division from materiality, showing how cyberspace hides its material allegiances. For scholars working within this intersection, the digital realm has an implicit politics of space, whether in the rhetoric of spatiality attached to the internet ("the cloud," "an information superhighway," "an ecosystem"), the representations of "nature" users interact with on the internet or in video games (Chang 2019), or the physical infrastructure that undergirds the internet (Hu 2015, Starosielski 2015). From undersea cables that stretch between continents, to data centers located in deserts and mountains, to satellites sent up into our atmosphere, to phones and computers, the internet grows from roots that stretch deep into the physical world (Carruth 2016: 368; Starosielski 2015). And these material roots create significant pollution: in 2013, the Centre for Energy-Efficient Telecommunications (CEET) estimated that the internet emitted some 830 million tonnes of carbon dioxide a year (the carbon emissions of a country like Germany) (CEET 2013: 2). Data centers, which keep the internet running, use energy from coal-firing plants and nuclear power stations. Bitcoin, a digital currency, produces the same amount of carbon dioxide a year as Sri Lanka or Jordan (Stoll, Klaaßen, and Gallersdörfer 2019).

Other digital ecocritics identify the ways cyberspace contributes to global warming and ecological collapse; I add to this work by arguing that perceptions of ecological collapse are at the root of cyberspace's creation. This essay explores the environmental and historical conditions that played a role in cyberspace's popularity in the 1980s and 1990s. In the first part of the essay, I argue that cyberspace's infinite virtual territory provided a solution to the apparent ecological crisis of the 1980s: the fear that the United States was running out of physical room to expand due to overdevelopment. In the 1980s, when global warming was just beginning to enter public consciousness, scientists and public commentators framed environmental crises through the lens of overdevelopment, overpopulation, and resource loss rather than the barely understood concept of climate change.[2] By discursively transforming the technology of cyberspace into an "electronic frontier," a term originated by technologist John Perry Barlow (1990), technologists, lobbyists, and journalists turned cyberspace into a solution for the crisis of overdevelopment and resource loss.[3] In a period when Americans felt detached from their own environment, cyberspace became both a new frontier for exploration and a so-called American space to which the white user belonged as an indigenous inhabitant. Cyberspace then allowed white Americans to set themselves as a racialized indigenous population threatened by colonization from the government and by Japanese corporations. This racialized frontier imaginary, in turn, shaped the forms of internet technologies and digital architecture throughout the 1990s, influencing the modern user's experience of the internet as a private space under their sovereign control.

In the second section of the essay, I hone into William Gibson's Sprawl trilogy (1984–1988), which critics often cite as the exemplary electronic frontier of cyberspace narratives. Counter to the critical consensus, I argue that the trilogy undercuts the frontier imaginary championed by early cyberspace adherents. If the fear of overdevelopment led to the desire for—and in some cases, the construction of—a new, virgin environment, the Sprawl trilogy dismantles this frontier imaginary by showing how cyberspace magnifies the environmental emergency of the physical world, acting as yet another invading agent. Paying close attention to the digital architecture of Gibson's cyberspace—the mechanisms of "jacking in," the physical impact of cyberspace, even the way content exists in cyberspace—reveals an infrastructure that treats hackers like resources, transforming them into part of the environment instead of promoting their independence.

Yet even Gibson's critique of the sovereign cyberspace user cannot help but mask the violence of cybercolonialism by privileging the white American user. This essay turns finally to Gibson's *Count Zero*

(1987), in which the white American hacker fears spatial "invasion" by (post)colonial countries. It is not just cyberspace that threatens the hacker's sovereignty but Haitian *loa* who emerge in the matrix and make the digital space their own. Gibson's undermining of the hacker's sovereignty through the loa plays into white settler's fears of colonization by racial others, particularly Black populations. Thus both the Sprawl trilogy and the discourse of the electronic frontier reinforce the ways the contemporary experience of cyberspace exists on the discursive patterns of colonialism. By framing the solution to an ecological emergency as an electronic frontier, technologists, journalists, and policymakers reinforced the idea that the only method of securing American freedom was through territorial conquest: a perceived environmental emergency could be fixed by taking over more (virtual) space.

In turn, cyberspace's roots in ecological crisis shed light on our inability to create a collective response to the current climate crisis. The individualism of the internet experience, developed as a response to both the environmental crisis and the fear of racial contamination, pushed internet users to see themselves as existing in individual worlds unattached to the larger world. In cyberspace, we are truly alone on the frontier, at a time when we can ill afford to be.

More Ecosystem than Machine

When John Perry Barlow (a journalist-rancher-turned-internet-activist), Mitch Kapor, and John Gilmore founded the Electronic Frontier Foundation (EFF), a foundation fighting for digital free speech, one of their aims was "to convey to both the public and the policy-makers metaphors which will illuminate the more general stake in liberating Cyberspace" (Barlow 1990). Thanks in part to the power of their discursive vision, the EFF had "substantial impact" on laws governing digital free speech and computing regulations in the 1990s (Turner 2006: 172–73). Using metaphors like the electronic frontier, the foundation shaped policymaker's and the public's vision of cyberspace, turning it from a technological issue to an issue of expanding and freeing territory. In this section, I argue that technologists, journalists, and policymakers constructed cyberspace as a solution to the environmental emergency of the 1980s, as an electronic frontier that would provide space to escape overdevelopment and renew (white) Americans' connection to the environment. By framing the technology of cyberspace through frontier rhetoric, technologists fought for an internet where users would have individual, private, and near-total control of their experience. For white Americans, the electronic frontier also provided a space to appropriate an experience of indigeneity. The discourse around cyberspace freedom sets up hackers as a racialized American population, threatened by Japanese colonization, rewriting the violent history of settler colonialism.

The frontier promoted by cyberspace adherents was not the historical frontier of the United States but rather a frontier born of myth and nostalgia. In his introductory essay to the 1994 textbook *Computerworld*, "Jack In, Young Pioneer!," Barlow captures the frontier nostalgia shared by his fellow cyberspace adherents: "I grew up resenting that the noble, essentially human, act of plunging off into unassayed wilderness, driven by nothing more rational than vague dissatisfaction and aspiration, would not be mine to undertake. It was the critical part of my inheritance which my forebears had spent." In Barlow's imaginary, the American landscape is "unassayed wilderness"—an untouched, empty land, free for white settlers to explore. Barlow asserts that exploration of the frontier constitutes a crucial part of his "inheritance." Of course, this myth erases the genocide of Indigenous people that created the "wilderness" Barlow's ancestors could explore and settle. Instead, Barlow and his fellow (white) Americans feel entitled to the experience of the wilderness through "inheritance." It was this mythological cyberspace, powerful in the imaginary if lacking in historical fact, that would become the basis for the electronic frontier discourse.

Cyberspace existed in the American imagination long before it existed as a mainstream technology. The cyberspace craze of the 1980s came not from a new invention but from literature. The term *cyberspace* came into widespread use when Gibson used it to describe the virtual world in his 1984 novel *Neuromancer* (5).[4] While virtual environments and virtual reality predate *Neuromancer*, they were accessible only to small groups of Americans—academics, researchers, and the military. *Neuromancer*'s popularity spread the idea of virtual reality to the American reading public, generating interest in the potentials of interconnected computer networks. Technologists took advantage of cyberspace's newfound popularity, using *Neuromancer*'s terminology and its portrayals of a cyberspatial infrastructure to inspire their own scientific innovations. In 1988, the cofounder of the virtual reality company Autodesk, John Walker (1988: 452), invoked Gibson's vision in a white paper, proposing a project that would produce "a doorway into cyberspace." Three years later, the manager of Autodesk's cyberspace project claimed his goal was to replicate *Neuromancer*'s deck and jack: "While

Gibson's vision is beyond the reach of today's technology, it is none-theless, today, possible to achieve many of the effects alluded. A number of companies and organisations are actively developing the essential elements of a cyberspace deck . . . these groups include NASA, University of North Carolina, University of Washington, Artificial Reality Corp., VPL Research, and Autodesk" (Walser 1991: 37). From public research institutions (NASA, public universities), to multinational companies (Autodesk), almost everyone used Gibson's work as a blueprint for cyberspace, a map into a world that no one had ever seen.

If the science fiction of the 1980s gave rise to the concept of cyberspace, it was the nonfiction of the 1970s that contributed to a feeling of ecological crisis: one in which Americans were running out of physical space for growth. In the spatial fix, David Harvey's (2001: 24) theory of the capitalistic desire to accumulate space, capitalist economies have an "insatiable drive to resolve [their] inner crisis tendencies by geographical expansion." Capitalism, for Harvey, can only survive by being "geographically expansionary" (25), a need that leads to crises because of the limited amount of space on the globe. The impossibility of geographical expansion came into public consciousness thanks to bestsellers like *The Population Bomb* (1968) by Paul R. and Anne H. Ehrlich and *Limits to Growth* (1972) by Donella Meadows, Jorgen Randers, and Dennis Meadows, which predicted that the earth's resources would no longer support the current rates of economic and population growth. As the limits of geographical expansion became clear, Americans believed they were on the brink of "what many believed to be an imminent ecological disaster . . . the earth as we know it might disappear" (Turner 2006: 120). With the end of capitalism nowhere in sight, the collapse of the environment seemed imminent.

Americans found a xenophobic outlet for their fear of overdevelopment: Japan's economic rise. In the early 1980s, the United States was the largest creditor nation in the world; by 1986, Japan had taken its place as the largest creditor nation, with the United States as the largest debtor nation (Heale 2009: 24). In his State of the State address in 1982, Governor Jerry Brown Jr. of California claimed that the United States was forming "a type of colonial relationship with Japan. We ship her raw materials, she ships us finished goods" (Brown quoted in Mathews 1982). The American business community feared the trade deficit meant that Japan would hold the United States hostage and become their "landlord and employer" (Heale 2009: 27). These commentaries mask anti-Japanese racism behind fear that Americans would lose control of their physical space by becoming mere renters of the

United States. For both Governor Jerry Brown Jr. and the American business community, Japan's economic ascendance over the United States threatened Americans' relationship to their country; the United States no longer belonged to Americans but to foreign "employer[s]." As Japan became the United States' economic rival, commentators channeled their anxiety into a fear that Japan would dispossess the United States of its land, alienating American citizens from the environment entirely.

In turn, technologists cast Americans' loss of control of their territory as an environmental emergency that threatened human independence: the shrinking of the world would force people to become deindividualized parts of a larger organism in order to survive. They honed into the infrastructure of the (Japanese) corporation, which technologists framed as an inhuman organism that assimilated workers into itself. Barlow (n.d.) explains the infrastructure of the corporation when talking about his time at Apple: "individual humans, regardless of station, are about as likely to 'run' larger corporations as coral polyps are likely to run reefs. I've been told Japanese management understands this a little better." In an interview, Barlow shifts the metaphor from coral to slime mold, arguing that "to say individuals are still individuals when they're acting in their corporate form is like saying slime mold is still a bunch of slime mold cells. We still have this Newtonian, causal, deterministic notion that organizations are machines" (Barlow and Kapor 1991: 47). For Barlow, the Japanese corporation is not a machine but an organism, a living creature that paradoxically dehumanizes all the workers within it. The corporate model remakes individual humans into smaller parts of its bodies, destroying individual needs to serve the interests of the colony.

For white Americans, these corporations promoted a loss of individuality through environmental collectivity. Stephen Hong Sohn (2008: 8) explains evolutions of Orientalism in the 1980s: "In traditional Orientalism, the East often is configured as backwards, anti-progressive, and primitive. In this respect, techno-Orientalism might suggest a different conception of the East, except for the fact that the very inhuman qualities projected onto Asian bodies create a dissonance with these alternative temporalities." The technological prowess of Japan did not wipe out the Orientalist configuration of the Japanese as primitive but rather cast them as an inhuman form of futurity (8). This inhuman futurity, for American technologists, took the form of the Japanese corporation, which transformed American individualism into environmental collectivity. Barlow's metaphorical organisms break the border between self and community: while slime mold and coral can exist as single cells or single plants, they come together in much larger organisms to survive or thrive (polyps into colonies, colonies into reefs, cells into amoeba into fruiting bodies). Slime mold clusters under conditions of emergency: when food is scarce, cellular slime mold joins into a slug-like creature, which crawls into a new place to become a fruiting body.

For others, the ecological crisis of the 1980s was imminent not because of the rise of Japan but because of the power of the American government. Charles A. Reich's *The Greening of America* (1970) portrays America as in "consciousness II," a state in which governments and corporate bureaucracies manage both people and nature through technologies of control and communication (67–70). The American environment was too controlled by the state to be a space of freedom and growth. "Cyberspace and the American Dream: A Magna Carta for the Knowledge Age," an influential polemic put out by the right-wing Progress & Freedom Foundation, also places fault in bureaucratic powers over nature, complaining that the "industrial age encouraged conformity and relied on standardization," a process that degraded "individual liberty" (Dyson et al. 1994). Bureaucracy and government had cut off so-called ordinary Americans—presumably, middle- and upper-class Americans—from the environment, alienating them from their land.

This fear that Americans could no longer access their own environment was fueled by the failure of the back-to-the-land movement. If for white Americans the rise of Japan and of bureaucratic oversight threatened to dispossess them of their environment, the failure of communalism seemed to prove that, even barring foreign interference, white Americans could not possess the environment at all. The period of the late 1960s and early 1970s saw the largest wave of communalism in American history, as groups of young, primarily white and upper-class Americans left the cities in droves to create settlements in rural America. Some 750,000 people settled into rural communes around this period (Miller 1999: xviii–xx). These back-to-landers sought both independence and a reconnection with nature, a desire for an authenticity rooted in a connection with the environment. As Fred Turner (2006: 74) points out, back-to-landers also evoked the imaginary of the frontier: "many communards saw themselves as latter-day cowboys and indians moving out onto the open plains in order to find a better life." Stewart Brand, whose magazine *Whole Earth Catalog* connected communalists, saw the communes as "Americans dwelling in the wilderness of changing eras . . . relearning to be natives from the most native Americans, the Indians, studying with new clarity the most ancient harmony of a shared land-heritage" (quoted in Turner 2006: 59). Despite his nod to Native Americans as the true holders of native knowledge, Brand hints at a further expropriation of Indigenous land by claiming that settlers have a "shared-land-heritage" with Indigenous people. Not surprisingly, communes represented further colonization, as urban dwellers encroached on the territory of rural Chicano and African American communities (Turner 2006: 77). Almost all rural communes had failed by the mid-1980s. Former urban dwellers struggled to farm the land, and funds dried up. If communalists desired a connection to the land, the failure of communalists showed how the land itself resisted such connection — and the frontier independence that came with it.

In reopening the space of the frontier, cyberspace provided a new environment for Americans to enter, an ecosystem that molded to the desires of its users and that stood outside the threats of invasion and overdevelopment. Critics widely acknowledge Barlow as the first person to use the term *cyberspace* to describe the intersection of telecommunications and computer networks that would become the public internet: in his first usage, Barlow called cyberspace "an electronic frontier." A few years later, Barlow (1994b) linked the need for cyberspace to nearly every point of the perceived 1980s environmental crisis: "We must seek our future in the virtual world because there is no economic room left in the physical one. Not only has all the good farmland been homesteaded long since, but nearly all the work one might do with his or her hands is now being done either by machines or by people from parts of the world where what's being considered a living wage is a lot less than you'd likely accept." In a few sentences, Barlow claims that the United States has no space, that bureaucracy takes up all remaining space ("no economic room"), that new farms are impossible, and that foreigners are taking away all remaining opportunities. From the commune problem (homesteading) to overdevelopment, Barlow covers them all — and presents cyberspace as the solution to every one of these manifold environmental issues. If capitalism survives through geographical expansion, cyberspace seems the ultimate solution to the spatial fix, because it still has plenty of "economic room" and infinite expansion. At the same time, cyberspace serves as a technological fix — the use of technology to solve a problem that, itself, is created by technology in a kind of infinite cycle of technological addiction. While the corporation evokes assimilation into the environment, cyberspace, with its vast swaths of emptiness, evokes the image of the American frontier—the mythological basis of America's rise to economic greatness.

By the mid-1990s, the concept of cyberspace as the American frontier dominated discussions of the emerging public internet. For journalists and technologists, cyberspace represented the frontier model brought into its ideal form: a place where spatial exploration unlocked the user's freedom. "Cyberspace and the American Dream: A Magna Carta for the Knowledge Age" cast cyberspace as a place to renew the frontier dream: "And as America continued to explore new frontiers . . . it consistently returned to this fundamental principle of rights, reaffirming, time after time, that power resides with the people. Cyberspace is the latest American frontier" (Dyson et al. 1994). Cyberspace represents a new opportunity to return to "this fundamental principle of rights"—for the libertarian "Magna Carta," those rights are individual freedom, decentralization, customization, and individuality. Even cyberspace's detractors bought into the frontier model. Ziauddin Sardar (1996: 33), for example, called cyberspace "nothing more than the 'virgin land' concept of the original exploration of the New World." While Sardar critiques the frontier model, pointing out that this new discourse erases the violent history of Indigenous genocide and conquest, he still sees the frontier as the correct lens through which to view cyberspace.

With the failure of physical communes and their promised freedom, new cyberspace adherents entered the electronic frontier as a way to extend the cowboy dream. Former commune dwellers of the 1960s and 1970s built and managed the Whole Earth 'Lectronic Link (WELL), one of the oldest and most popular virtual communities. Brand, a proponent of communalism, started the WELL, named after his communalist catalogue (the *Whole Earth Catalog*); managers included Cliff Figallo and John Coates, former members of a Tennessee commune (Turner 2006: 142). The WELL, in turn, hosted such cyberspace influences as Barlow, Howard Rheingold, and Kapor—journalists and entrepreneurs who shaped public discourse around cyberspace. For these new cyberspace adherents, the hacker was the direct descendant of the cowboy. Writers cast hackers in cyberspace as "homesteaders" (Elmer-Dewitt and Jackson 1994), "mavericks" (Leary 1988: 253), "cyber-tribal hunter-gatherers" (Barlow 1994a), "mountain men," "desperadoes," "vigilantes," and "pioneers" (Barlow and Kapor 1991: 47–49)—all the true inheritors of the American frontier. Of course, much as the electronic frontier was a mythological construction that erased historical fact, the cowboy dream ignored the real cowboys of the United States, few of which were explorers, many of whom were Black and Indigenous.

The casting of the hacker-as-cowboy shaped modern user's experience of the internet as a space of individual, private, and controlled explorations. To be a cowboy in cyberspace, to inhabit the frontier values of self-governance and exploration, required individual and private access to the internet. Cowboy-hackers promoted "virtualization": the method of "constructing a simulated environment that both allows a user unprecedented freedom (it seems as if she has control over an entire virtual environment) and restricts that user from 'leaking' or contaminating the data of other users" (Hu 2015: 61). Virtualization was never a guarantee: in the 1960s, digital communication networks that predated the internet were envisioned as a public utility with no private individual users (Hu 2015: 52–64). Now, as Wendy Hui Kyong Chun (2011: 102) explains, the internet creates a feeling of an individual "'you' as the sovereign subject, 'you' as the decider." When a user goes on the internet, in other words, they seem in sole control of their explorations, their movements, the content they upload (or view); the territory is theirs to take. This cornerstone of contemporary internet infrastructure is a product of WELL and EFF's advocacy, their discursive transformation of cyberspace from a public infrastructure to a private space of exploration—the "inheritance" of the frontier.

If the cowboy stands for US freedom, the sovereign subject in control of their explorations, Japan comes to stand for the assimilationist racial other. The specter of Japan haunts calls for internet freedom through virtualization, the threat of assimilation that could destroy the "frontier" freedom of cyberspace. Tung-Hui Hu (2015: 88) argues that "the shadow of the racialized Other still resurfaces in calls for internet freedom," through the struggle between the liberal (white) society's desire for freedom and the enemy they create—the assimilationist racial others who desire totalitarianism. For Chun (2006: 187), "cyberspace appears to be a Western frontier in which US ingenuity wins over Japanese corporate assimilation, for cyberspace allows for piracy and autonomy. In stark contrast to those working for seemingly omnipotent zaibatsu, for whom power is gained through 'gradual and willing accommodation of the machine, the system, the parent organism,' the meatless console cowboy stands as an individual talent." Barlow and other EFF promoters cast the Japanese as the corporation, the machine, the reef—institutions or organisms in which individuals could have no freedom, a stark contrast to the maverick (white) cyberspace cowboys. As a response to America's fear of a rising Japan, cyberspace stood for the return of white American dominance and white American embodiment—the individual body—in a white American environment.

For early adherents of cyberspace, the very infrastructure of cyberspace—or lack thereof—made hierarchical governance impossible. In science and technology studies (STS), scholars of internet infrastructure explain that the internet functions thanks to a series of digital systems and processes like Internet Protocol (IP) addresses, autonomous system numbers (ASNs), and physical-digital interfaces like telecommunication services, service providers, and electricity. This technical architecture remains out of public view, below the level of content (DeNardis 2012: 721). But while contemporary STS scholars argue that politics existed at the level of digital and physical architecture, not just content, during the early internet era, technologists saw digital architecture as apolitical (DeNardis 2012). Politics played out at the level of content—and since digital content had no physical existence, it could not be controlled by governments or corporations. Thus, while the physical American frontier came under the rule of the government, Barlow (1996) argues that cyberspace works differently: it lies outside of physical space. In "A Declaration of the Independence of Cyberspace" (1996), he tells governments that: "Cyberspace does not lie within your borders. Do not think that you can build it, as though it were a public construction project. You cannot. It is an act of nature and it grows itself through our collective actions . . . legal concepts of property, expression, identity, movement, and context do not apply to us. They are all based on matter, and there is no matter here." For Barlow, the physical and digital infrastructure of cyberspace (which was built and controlled by corporations and governments) is much less important than the space itself. Barlow's statement that cyberspace cannot be built "as though it were a public construction project . . . it grows itself through our collective actions," makes it clear that cyberspace's political "material" is its contents—the writings and communities generated by cyberspace's early entrants. In this division of infrastructure and space, governments could control digital infrastructure, but because content itself had no physical matter, cyberspace still promoted an anarchic, unhierarchized existence.

Electronic frontier proponents worked to dismiss cyberspace as a form of infrastructure by casting it as an ecosystem, an environment, or a frontier: something that no one built but that grew out of hierarchical control. The Bill Clinton/Al Gore administration of 1992–2000 termed the internet an "information superhighway," tying cyberspace to government infrastructure projects—and their regulations. When Barlow tells the American government, "Do not think that you can build [cyberspace], as though it were a public construction project," he challenges the government's vision of cyberspace as a "public project," or another form of mass infrastructure. Those sympathetic to Barlow's vision claimed that cyberspace grew "naturally" out of the control of bureaucracies. The "Magna Carta" claims cyberspace is "more ecosystem than machine" and a "bioelectronic environment" (Dyson et al. 1994). The report argues that the rise of cyberspace would destroy all hierarchies: "it shapes new codes of behavior that move each organism and institution—family, neighborhood, church group, company, government, nation—inexorably beyond standardization and centralization, as well as beyond the materialist's obsession with energy, money and control . . . It also spells the death of the central institutional paradigm of modern life, the bureaucratic organization" (Dyson et al. 1994). While the "Magna Carta" recognizes the current governmental control over cyberspace, for the writers, cyberspace's lack of mass ("demassification") would undermine standardization, bureaucracy, and centralization. In this frontier imaginary, the ecosystem of cyberspace destroys government infrastructure just by existing.

The casting of cyberspace as an environment rather than an infrastructure did not just revive the American frontier: it gave white Americans a space in which to appropriate indigeneity. Unlike the American landscape, which belonged to white Americans by dint of genocide and Native displacement, in cyberspace, white Americans could be the first inhabitants. As a new environment, "more ecosystem than machine," it had no prior allegiances to anyone except the (American) users (Dyson et al. 1994). Technologists often found themselves slipping into the language of indigeneity, calling early residents of cyberspace "natives" and "primitives" (Barlow 1994a) and comparing the American government to a colonizer bent on civilizing their anarchic utopia (Barlow 1996). It is no coincidence that Barlow, a major promoter of the digital native, was a lyricist for the Grateful Dead, a band that used and appropriated Native American iconography and motifs in their sets and to advertise their music. As Philip J. Deloria (1998: 183) argues, the appropriation of indigeneity by Americans "offered a deep, authentic, aboriginal Americanness . . . to play Indian has been to connect with a real Self, both collective and individual, and there was no better way to find such reassurance." In cyberspace, white

Americans—just white Americans—could be both cowboy and Native, colonizer and Indigenous. The white appropriation of indigeneity allowed for an authentic connection to an Americanized environment (the frontier) outside of the communal assimilation of corporations and institutions.

By turning cyberspace into a frontier, early cyberspace adopters rewrote the history of colonization, turning American hackers into cyberspace's native dwellers and the government into the colonial invader. For cyberspace adherents, indigeneity was a learned position rather than an identity. Gurney Norman (quoted in Turner 2006: 87), who wrote for the *Whole Earth Catalog*, explained to the predominantly white readers of the *Catalog* that not all "indians" were Native Americans: "today, they are joined by others who qualify as 'Indians' of a sort, by virtue of their skills which allow them to function as teachers, as shamen [*sic*], as knowers of The Way . . . certain farmers and artisans, aborigines of a kind, native to their place, there on the land to be learned from." In this construction, indigeneity comes from one's connection to the land and can be learned—from Native Americans but also from shamans, teachers, and farmers.

This learned form of indigeneity, in turn, translated into cyberspace. Barlow (1994a) explains how cyberspace dwellers are natives in an early piece on cyberspace and copyright: "soon most information will be generated collaboratively by the cyber-tribal hunter-gatherers of cyberspace. Our arrogant legal dismissal of the rights of 'primitives' will soon return to haunt us." For Barlow, more and more people will become "cyber-tribal hunter-gatherers" of cyberspace and take on the position of "primitive" thanks to their skills in telecommunication. Once Americans enter cyberspace, the government turns against them: "the government is preparing to place this new frontier under the rule of law. Whether the pioneers already there want it or not" (Barlow 1994b). David Gans of *Mondo 2000* compared the struggle for telecommunication freedom to "the struggles faced by southern blacks" before the "emancipation proclamation" (Barlow and Kapor 1991). These constructions problematically rewrite American history, with the white American hacker set up as either the racialized Indigenous population under threat of colonization, or the enslaved Black person deprived of freedom by an unjust government.

The technologist construction of the cyberspace hacker as a racialized population under threat from an oppressive government—or the oppressive Japanese corporation—reinforced the need for virtualization, the process that creates the private internet user. Virtualization serves not just as a solution to the environmental emergency by giving hackers a new frontier to explore, but it also appears to give the hackers freedom from unjust institutions that attempt to sublimate their individuality. The architecture of cyberspace allows the hacker to enter an untouched, new world as a perfectly sovereign individual, with the crisis of freedom and environment neatly solved through a virtual utopia.

Thirty years after the public opening of the internet, we are still looking for the new technological frontier, the space that solves the crisis of individuality and environment at the same time. The hackers and technologists who wrote the blueprint of cyberspace may not have created their vision of a new, untouched frontier in cyberspace, but the utopian impulse that seeks a technological resolution to the environmental crisis remains in the present day. Every time a new technological fix appears, it recreates the technologists' search for a virgin land whose freedom from pollution is created through technological means. It is not a coincidence that most of these solutions require a "new space" to explore and conquer, whether that be the ocean, a new form of virtualization, or the ground deep beneath our feet. Yes, cyberspace was not the promised frontier—but somewhere, out there, is the technical innovation that will simultaneously free us from climate disaster and allow us to keep a contemporary capitalist lifestyle. Whether it will be using solar shields to deflect heat away from the earth, or storing carbon deep inside the ocean, or replacing some of the ocean's carbonic acid with hydrochloric acid to speed up ocean storage of carbon dioxide, or just using more data centers—the new frontier is out there, if only we can build it.

But of course, every new iteration of the technological frontier builds on top of its ancestors, a layered graveyard of territories and utopias, colonial hardware stacked atop more colonial hardware. Undersea cables that tie the internet together from continent to continent are built atop undersea telegraph lines (Hu 2015: 36–38; Starosielski 2015: 12–15). The supposedly wireless internet is wired onto the infrastructure of nineteenth-century imperialism. Data centers are built out of Cold War bunkers (Hu 2015). As Starosielski (2015: 31) points out in her comprehensive study of undersea cables, "the geography of telegraph routes in the late nineteenth century followed transportation and trade routes, many of which had been pioneered by British colonial investment and served to support existing networks of global business." The next technological fix, too, will build on these now-ancient monuments: geoengineering will almost certainly use the technology of cyberspace in the promise of a pollution-free world, even as the polluting fossils of ancient telegraph lines, Cold War bunkers, and data centers stand as an omen of their future.

A Polyp in a Coral Reef

Perhaps the first link between the cowboy and cyberspace appears in William Gibson's cult novel, *Neuromancer* (1984), when the protagonist muses that "he was no console man, no cyberspace cowboy" (5). The "console cowboys" (28) and "cyberspace cowboy[s]" (5) of the Sprawl trilogy (*Neuromancer*, *Count Zero*, and *Mona Lisa Overdrive*) helped forge the imaginative link between the white American cowboy myth and virtual reality, the act of hacking as exploration. Barlow (n.d.) himself endorsed *Neuromancer* as a pioneer of the electronic frontier; Barlow credited *Neuromancer* with making him realize that the internet was a "cyberspace," or "a 21st Century digitally created landscape of information and mind." The trilogy's focus on the infinite space of cyberspace, its love of Western imagery, and its techno-Orientalism made it the emblem of technologists in the 1980s and 1990s (Roh, Huang, and Niu 2015: 2–9).

Yet despite Sprawl's emblematic role in constructions of the electronic frontier, in Gibson's trilogy, cyberspace functions as an invader rather than a liberator. The technological solution to the environmental emergency of the 1980s was the construction of a sovereign user and an environment over which they could reign: the hacker/cowboy and their electronic frontier. In *Neuromancer* and *Count Zero*, the digital architecture of cyberspace functions as an ecosystem, the virtual environment that hackers so craved. But this ecosystem is not a frontier space; instead, it functions as an assimilationist ecosystem that destroys the sovereignty of the user, dismantles virtualization, and creates a form of environmental collectivity much like the hive. Unfortunately, the hacker's assimilation into cyberspace does not mean Sprawl opts out of cybercolonialism: the final part of this section explores the Haitian loa of *Count Zero*, who Gibson characterizes as both avatars and invaders of cyberspace. Even in Sprawl's critique of the cyber-frontier, Gibson privileges the experience of the white American hacker by casting racial others as a threat, mimicking the technologist fear of Japanese colonization and incorporation through the loa.

One of the reasons that critics do not apply an infrastructural lens to the Sprawl trilogy is that the books contain very little physical infrastructure. But Sprawl does have digital architecture: the infrastructure within cyberspace itself that allows virtual reality to function. The digital architecture of the internet includes physical-digital interfaces (telecommunication services, service providers, electricity) and digital processes that allow people to move between different parts of the internet. As Laura DeNardis (2012: 720–21) argues in her work on internet governance, focusing just on the contents and the owners of the internet often causes us to miss the way that the internet's technical architecture functions as both a form of government and as "arrangements of power." In other words, cyberspatial infrastructure is, in and of itself, a political entity, beyond cyberspatial content. While DeNardis and her colleagues in STS work on the internet, which opened to the public years after the Sprawl trilogy (DeNardis and Musiani 2016: 3–12), her observations are evident in the fictional cyberspace of the Sprawl trilogy: from the jack that lets hackers enter the matrix to the system of virtualization that keeps hackers from seeing each other in cyberspace.

The most visible part of cyberspace's technical architecture in the Sprawl trilogy is "jacking in": the process of a hacker entering cyberspace. "Jacking in" undercuts the disembodiment hackers believe they access in cyberspace: the jack gives cyberspace access to the hacker's body even as they enter into the disembodied matrix. To "jack in" involves plugging the cyberspace deck into a neural interface; it establishes a physical connection between the plug of the deck (an early model computer) and the "jack" of the hacker's neural interface (Gibson 1987: 17). All attempts to enter cyberspace involve a corresponding invasion of the body, a penetration of cyberspace into the hacker's nervous system (Gibson 1987: 3; Gibson 1984: 54). Chun (2006: 3) explains this dynamic as it relates to the internet: "The moment you 'jack in' (for networked Macs and Windows machines, the moment you turn on your computer), your Ethernet card participates in an incessant "dialogue" with other networked machines . . . your computer sends information, such as your Internet Protocol (IP) address, browser type, language preference, and userdomain," which in turn contains information about your physical location. The dialogue Chun describes involves an exchange of information between computer and internet that occurs below the user's awareness. The internet draws information from the user's machine; the user perceives only the information they consent to put out, not the information leaking out from their machine to the internet. In much the same fashion, the "jacked in" hacker in *Neuromancer* and *Count Zero* only perceives their body entering cyberspace—not cyberspace entering their body and leaking information from their nervous system into the digital world.

By invading the hacker's nervous system, the jack dismantles the hacker's sovereignty, undermining the "cowboy" independence sought by hackers and technologists alike. While the electronic frontier promoted the hacker as a cowboy, a figure in harmony with and in control of the landscape, cyberspace in Sprawl removes the hacker's sovereign control over territory by transforming them into an ecosystem. When surgeons repair Case's—the protagonist and main console cowboy in *Neuromancer*—connection to cyberspace, "Black fire found the branching tributaries of the nerves, pain beyond anything to which the name of pain is given . . . And Ratz was there, and Linda Lee, Wage and Lonny Zone, a hundred faces from the neon forest, sailors and hustlers and whores, where the sky is poisoned silver, beyond the chainlink and the prison of the skull . . . the sky faded from hissing static to the noncolor of the matrix" (Gibson 1984: 31). If the internet circulates users' information from their machines (Chun 2006: 11), cyberspace circulates hackers' memories from the roots of their nervous systems, making them part of the cyberspatial landscape. The nervous system transmits information about the body's perception of the world: by joining the nervous system and its neural pathways to cyberspace, Case and his surgeons route his embodied experiences straight down the neural circuits into cyberspace itself. Case's memories follow the "tributaries," the neural pathways, into the virtual world, transmitting fragments of Case's self beyond the digital border. Worse, contact with cyberspace turns Case into an ecosystem, his nervous system and brain becoming a landscape of tributaries and poisoned skies.

In some ways, the digital architecture of cyberspace resembles the very hive-like structure of environmental collectivity that hackers seek to escape. Like a hive, digital infrastructure prioritizes a distributed, multiplied consciousness over autonomous subjectivity. In the last scene of *Neuromancer*, Case sees "three figures, tiny, impossible, who stood at the very edge of one of the vast steps of data . . . Linda still wore his jacket; she waved, as he passed. But the third figure, close behind her, arm wound around her shoulder, was himself. Somewhere, very close, the laugh that wasn't laughter" (Gibson 1984: 270–71). The second and third figure are Linda Lee, killed in the first thirty pages of the novel, and Case himself. The doubles of Case and Linda Lee are creations of Case's memories, fragments of the past animated by the digital architecture that accessed Case's neural system. Rather than experiencing freedom by entering cyberspace, Case is now possessed by digital infrastructure, transformed by cyberspace into a whole new entity under its direct control. Through the mechanism of the jack and the invasion of the hacker's nervous system, cyberspace remakes individuals into small pieces of itself. Instead of the cowboy, they become a cell in slime mold, or a polyp in a coral reef, their individuality subsumed to the interests of a larger organism.

The clearest sign that cyberspace in Sprawl dismantles cowboy sovereignty is the slow dissolving of virtualization over the course of the trilogy. Virtualization is the process that creates the private user by separating each user's data from another's; it is an ideology as well as a process, a politics of treating cyberspace as a private and individual space of exploration (the frontier) rather than a collaborative public place (Hu 2015: 52–64). The hackers experience near-perfect virtualization in the matrix of Sprawl: when a hacker enters the matrix, they always appear to be the only user and never see any other hackers, even if others are in the same location. They encounter the matrix as the sole and sovereign explorer. But the heist plot in *Neuromancer* aims to merge private data—and private users. The mysterious Wintermute (an AI residing in cyberspace) hires Case to help Wintermute join with his other half, an AI named Neuromancer. The heist sees cyberspace move from a space of privacy and strict boundaries to a space of data merging and collaboration—a dismantling of the virtualization that kept the AIs apart. The end of *Neuromancer* and the plot of *Count Zero* have hackers encounter other agencies in cyberspace. In *Count Zero*, Finn, a seller of stolen goods, remarks, "Yeah, there's things out there [in cyberspace]. Ghosts, voices, why not? Oceans had mermaids, all that shit, and we had a sea of silicon, see? Sure, it's just a tailored hallucination we all agreed to have, cyberspace, but anyone who jacks in there knows, fucking *knows*, it's a whole universe. And every year, it gets more crowded, sounds like" (Gibson 1987: 119). Cyberspace replicates the overcrowding that drove hackers into the virtual while breaking down the hacker's privacy: they now meet other users, other agencies, even crowds, instead of experiencing the matrix as their private frontier.

These crowds indicate that cyberspace produces its own natives, disrupting the hacker's claim to indigeneity. If much of the technologist discourse on cyberspace casts early computer users as "natives" and "aborigin[es]" (Barlow 1994a), in Sprawl, the infrastructure gives rise to new agencies, ones that block human attempts to become native within the matrix. These agencies emerge first in *Neuromancer*, with the AI seeking freedom from their creators, and with the doppel-gangers in the matrix—Case's double, Linda Lee—conscious frag-ments of human memory who, unlike hackers, can navigate the digi-tal landscape as embodied digital creations. The most mysterious and powerful of cyberspace's agencies emerge in *Count Zero*: the loa, Hai-tian gods who remake cyberspace in their own image. Unlike the AI, who are human creations, or the doppelgangers, who come from cyber-space, the loa's provenance is altogether unknowable. White hackers interpolate them as viruses, as evolutions of AIs, and even as ghosts. The hackers' inability to trace the provenance of the loa speaks both to the way digital architecture works below the level of content for users and to cyberspace's separation from the intentions of users. Although cyberspace seems to have no prior allegiance except to the (American) user/hacker, the presence of these (Caribbean) agencies puts the hacker once more in the position of the colonizer rather than the native, the intruder rather than the welcome explorer.

That the new, native agencies of the Sprawl trilogy's cyberspace are from the Caribbean is crucial, because in the 1980s, US institutions cast Caribbean people—and specifically Haitians—as agents corrod-ing the United States from the inside. The American fear of Caribbean people crystallized around Caribbean immigrants to America: US offi-cials feared that the Caribbean refugee crisis in the summer of 1980 would lead to the United States being overwhelmed by asylum seek-ers (Lindskoog 2018). As Haitian refugees kept coming, the Reagan administration changed their status to "economic migrants," detaining them without bond. In the meantime, the American medical commu-nity and the Reagan administration scapegoated Haitians as the prove-nance of AIDS in the United States (Farmer 2006: 2). Starting in March 1983, the Centers for Disease Control and Prevention grouped Hai-tians with three other groups (homosexuals, hemophiliacs, and heroin-users) at "high-risk" of AIDS, and soon after, members of the group were collectively known as the "4H club" (Farmer 2006: 211). Scien-tists suggested that Haitians were responsible for bringing AIDS to the shores of the United States through the mechanism of Vodou (the very religion associated with loa). In 1983, the *Annals of Internal Medicine* published a report from physicians who claimed that "it seems reasonable to consider voodoo practices a cause of the syn-drome [AIDS]" (Moses and Moses 1983: 565). White Americans feared a transformatory infection from Haitians: that the immigrant population would replicate their Caribbean landscapes in America through numbers and disease. Nor did the US government fear only

Caribbean migrants: during the 1980s, the Reagan administration backed Haiti's dictatorship, used military force to decimate Grenada's Marxist independence movement, and provided neoliberal parties in Jamaica with CIA training and weapons to contain what it perceived as the threat of communism so close to the United States.[5]

The possession of hackers by the loa in *Count Zero* represents the possession of both white American bodies and US territory by a formerly colonized space: Haiti. The "horse," a figure in Haitian Vodou practice, exemplifies this transformation of hackers by a Caribbeanized cyberspace. The horse, in Gibson's gloss of Haitian Vodou, mediates between humans and the loa; the loa "ride" humans, who become their "horses" (Gibson 1987: 113). The loa possess many characters, but the main horse in *Count Zero* is Jackie, a Haitian American mambo who serves as the horse to Danbala, the serpent loa who acts as the creator of all life. At the novel's climax, Bobby (a white American hacker and the titular Count Zero) confronts the antagonist, Virek, and Baron Samedi, another loa, possesses him: "'My name,' a voice said, and Bobby wanted to scream when he realized it came from his own mouth, 'is Samedi, and you have slain my cousin's horse . . .'" (Gibson 1987: 293). In an earlier part of the plot, Papa Legba, the master of the roads, takes over the body of a young American girl named Angie Mitchell so he can propose a deal to her bodyguard, Turner: "This child for my horse, that she may move among the towns of men" (233). It is hard to imagine anything more opposite to the kind of individual transcendence hackers seek in cyberspace than total possession by a god. The loa ride hackers, treating them as pawns for their plans—Samedi uses Bobby because Virek killed Jackie, "my cousin's horse"—their selves sublimated to the desires of entities they do not understand.

In a reproduction of the technologists' fears of Japanese colonization and incorporation, the hackers become avatars of Caribbean cyberspace, transforming the United States into a copy of the Caribbean from the inside out. Angie's father, for example, alters Angie's nervous system so she can access the matrix without the use of a deck. Her constant connection to the matrix transforms Angie into an unwitting horse for Ezili Freda, Papa Legba, and Baron Samedi, three loa inhabitants of cyberspace. As a horse for three cyberspatial loa, Angie becomes a resource for both cyberspace and the Haitian gods. The loa possess Angie's body whenever they want to access the physical world, as in this moment when Legba directs the mercenary Turner through the city: "Now that same flood of language, a soft fast rattle of something that might have been patois French . . . the deep muscles of her face had contorted into a mask [Turner] didn't know. 'Who are you?' 'I am the Lord of the Roads'" (233). Angie is not just an agent of Legba's; she seems transformed by Legba's presence, turned into a version of him. Legba's possession makes Angie into a person who can speak "patois French" (*Kreyòl*), whose features are unrecognizable. Angie asks Turner over and over if she's sick; several people diagnose her as cancerous. For Turner and the rest of the American hacker community, Angie's body becomes a piece of Haiti, bringing the Caribbean into the United States and contaminating the territory around her. In this case the fear is one of hijack—that a virus (or cancer cell) has taken over the (human/cybernetic) system and is transforming it to look like a copy of the virus, destroying the host in the process.

This hijack helps explain what happens to the Wig, a hacker who steals a fortune from African governments. After suffering a mental breakdown, the Wig becomes convinced that God lives in cyberspace and implants a neurological jack in his brain so he can hear "the voice of God," which tells him he needs to "get up the gravity well, God's up there" (Gibson 1987: 156–57). Once in outer space, the Wig starts obeying the commands of "God" by sending technology infected by the loa back to earth-bound hackers. Although the Wig begins his life as a user of cyberspace, plundering the virtual territory for resources, the loa assimilate him into their interests, control his movements, and use him to bring more people into contact with the loa and cyberspace. Angie and the Wig's bodies become avatars of the Haitian loa in the physical world, the first steps of a larger transformation of the American material landscape.

The deeper into *Count Zero* we read, the more Gibson makes cyberspace's threat of absorption racialized, a threat of invasion by Black Caribbean space. On the one hand, both *Neuromancer* and *Count Zero* disrupt the hacker's experience of the matrix as a private frontier, portraying cyberspace as a space that dissolves the user's experience of individuality. Through the mechanism of the jack and cyberspace's technical architecture, the hacker becomes a cell in a slime mold, a polyp in a coral reef, their interests and individuality subsumed to the interests of the larger ecosystem of cyberspace. But while Sprawl portrays the impossibility of escaping overdevelopment through cyberspace, it routes this impossibility through the specter of racial contamination. In a parallel to the technologist fear of the Japanese corporation, which had no room for the individual worker, the Caribbeanized cyberspace undercuts the private cyberspace user. While for the technologists who came after Gibson, cyberspace allowed for an escape from overdevelopment into the pleasures of the frontier, in Sprawl, cyberspace created the threat of invasion by (post)colonial countries. Both Sprawl and its descendants, in turn, privilege the experience of the white American user in the cyberspace frontier and mask the violence of colonialism that allows the frontier to exist.

Collective Immune System

As a means of conclusion, I want to consider one of the material consequences of cyberspace's frontier mythos: the contemporary exclusion of Native Americans from the internet. Constructed as a space in which white Americans could experience indigeneity, the digital frontier made no space for the experience—or presence—of Indigenous Americans, whose real-world experience frontier discourse erased. Fear of absorption into the racialized other has created a situation where racial others are often totally pushed out of cyberspace. For the white hackers of the 1980s and 1990s, Native people were inspirations for frontier mythos, but they were not comrades in cyberspatial exploration. The technolibertarians, who envisioned the cyberfrontier as a space of exploration for all, never considered the physical infrastructure necessary to connect Native people to the digital frontier. That dynamic remains true thirty years later. As of this writing, people who live on rural tribal lands have the lowest rates of internet access in the United States: 68 percent lack fast in-home internet (Diggins, Welzbacker, Shinner 2020). As many as 1.5 million people living on tribal lands cannot access internet services (Tabacco 2020). This digital divide has worsened during the coronavirus pandemic (which disproportionately affects the Native population of the United States). In a time when physical distance is critical, tribal residents lack access to education, public health information, and telemedicine, and tribal leaders struggle to communicate vital information (Estus 2020). While Gibson reveals the mechanism of the jack as restrictive, it is worth considering that most people on tribal lands lack any access to modern forms of the jack—wireless, high-speed internet—at all.

Settler colonialism excludes tribal lands and Native people from the internet. In "Jack In, Young Pioneer!," which gives this piece its title, Barlow (1994b) appropriates the Indigenous position, claiming that "the electronic frontier also differs from its predecessor in that setting up reservations is not likely to suffice for corralling the natives." Setting aside Barlow's casual racism ("corralling" is, at best, a dismissive way of explaining the genocide of the United States' Indigenous population), what Barlow ignores is the power of the reservation system on cyberspace. Reservations did—and continue to—stop people on tribal lands from accessing the internet. Because of federal Reservation Era policies set up in the late 1800s, the Federal Communications Commission (FCC) controls tribal spectrum resources (Tabacco 2020). Spectrum refers to the radio frequencies that the internet travels through—4G internet goes through the 1800MHz frequency, for example. Even when those frequencies travel through tribal lands, the FCC controls them and can sell them to the highest bidder (*Native Business* Staff 2020). In other words, people on tribal lands do not control their own cyberspatial resources. "Setting up reservations" always affected Indigenous people's access to the internet by stopping them from accessing the infrastructure they need to use it. But because Barlow constructed the electronic frontier as detached from the physical world, he and his allies made it easy to ignore the colonial consequences of cyberspace. For the Native people of America, the victims of the first frontier, that history remains much harder to ignore. While the cowboy-hacker helped create the sense of privacy that many contemporary users of the internet now enjoy, many more are on the outside of that virtual bubble.

This now-ubiquitous experience of virtualization, which prevents users from "leaking" into other users, trains internet users to experience themselves as moving through private, individual worlds that do not affect other people. In turn, the individualism of the internet experience, developed as a response to both the environmental crisis and the fear of racial contamination, restricts us from a collective response to the current climate crisis. The internet user cannot see the Native people outside the virtual bubble, nor can they see the pollution that their use of cyberspace creates.

It was the simultaneous white American fear of absorption into a racialized other—the Japanese corporation, the Black Caribbean immigrant, the postcolonial nation—and fascination with the freedom of indigeneity that produced the virtual bubble in the first place. The individual user of cyberspace, a being theoretically immune from both environmental and racial contamination, is a direct product of the environmental crisis of the 1980s and 1990s. Now cyberspace restricts us from a collective response to the climate crisis, locking some outside of the virtual bubble while it prevents those of us inside from seeing our effects on each other. In Barlow's metaphor of the corporation-as-slime-mold, he explains that "when [slime mold] decides that it wants to cover some country because conditions are changing . . . all the local slime molds get together and create an organism . . . it's not a one-celled animal any more, it's a multicellular organism, and it goes some-place and then it devos. It goes back down to its original constituents" (Barlow and Kapor 1991). Slime-mold cells aggregate together dur-ing conditions of environmental precarity: the cells are, at least on some level, aware that environmental conditions affect each other, and they create a collective response. But this is exactly the kind of collectivism that technologists built cyberspace's individualistic frontier to eschew. Instead, internet users see themselves less like slime mold and more like the console cowboy of Sprawl: a sovereign power untouchable by the environment and users around them. Yet as Sprawl demonstrates, every console cowboy is enmeshed in their environment, both cyberspatial and physical. The only difference between the internet user and the slime mold is that, unlike the slime mold, the internet user has lost their ability to aggregate together under conditions of environmental catastrophe. In 1991, Barlow argued that people joined corporations for safety: "You give up your mind for the benefit of the collective immune system that will protect you against the slings and arrows of individual fortune" (Barlow and Kapor 1991). At their most utopian, cyberspace's adherents hoped to create a space where such a collective immune response would no longer be needed, where it was safe to be an individual again. Instead, they created a space where such a collective immune response was no longer possible.

Suzanne F. Boswell is a PhD candidate at Rutgers University. Her research looks at the intersection of science fiction, environmental humanities, and Caribbean literature to uncover the relationship between tropical environments and twentieth-century lit-erary constructions of futurity. Her writing has appeared in *Paradoxa* and *Extrapola-tion.* For 2020–2021, she is an AAUW American Dissertation Fellow.

Notes

For their valuable feedback and encouragement on multiple drafts of this essay, I would like to thank Danielle Allor, Andrew Goldstone, Ariel Martino, Samuel Steinbock-Pratt, Michelle Stephens, and Erik Wade. For funding my archival research into cyberspace, I am grateful to the R. D. Mullen Award and the Mellon Foundation. And for time, solidarity, and encouragement when I was revising this work during the pandemic, I am grateful to the West Brooklyn Waterfront Mutual Aid community.

1   *Cyberspace* is a catch-all phrase used to refer to the environment in which communications over linked computer networks take place—whether virtual realities that occur in computers, the internet, or fictional digital realms.

2   The first use of the term *global warming* in a scientific paper came in 1975 (Broecker 1975: 460). The first intergovernmental panel on climate change did not meet until 1988. On the other hand, books like *The Population Bomb* (1968) by Paul R. and Anne H. Ehrlich, *The Greening of America* (1970) by Charles A. Reich, *The Closing Circle* (1971) by Barry Commoner, and the 1972 report *Limits to Growth* by Donella Meadows, Jorgen Randers, and Dennis Meadows—all of which discussed overdevelopment and resource loss—were bestsellers in the 1970s and 1980s.

3   Barlow first used of the term *electronic frontier* to describe cyberspace in the 1990 essay "Crime and Puzzlement." Barlow's frontier rhetoric echoed throughout 1990s writings, from journalism to scholarship. Often the rhetoric of the electronic frontier was used explicitly by scholars—see, for example, Sardar 1996. Politicians and lobbyists also echoed Barlow's frontier myth: see, for example, Dyson et al. 1994.

4   Gibson in fact first used the term *cyberspace* in his short story "Burning Chrome" in 1982, but the term did not gain widespread popularity until *Neuromancer*.

5   In addition to the inflection point of Haitian migration in the 1980s, Haiti holds particular status in any imaginary of the frontier or freedom, since it stands as the first country to ban enslavement. While Haiti's war of independence from France was partially inspired by the United States' earlier revolution, the United States government rejected Haiti, fearing that Haiti's successful revolt would inspire enslaved people in the United States to follow Black Haitians' example.

References

Barlow, John Perry. 1990. "Crime and Puzzlement." *Electronic Frontier Foundation*, June 8. https://www.eff.org/pages/crime-and-puzzlement.

Barlow, John Perry. 1994a. "The Economy of Ideas." *Wired*, March 1. https://www.wired.com/1994/03/economy-ideas/.

Barlow, John Perry. 1994b. "Jack In, Young Pioneer!" *Electronic Frontier Foundation*, August 11. https://www.eff.org/pages/jack-young-pioneer.

Barlow, John Perry. 1996. "A Declaration of the Independence of Cyberspace." *Electronic Frontier Foundation*, February 8. https://www.eff.org/cyberspace-independence.

Barlow, John Perry. n.d. "Leaving the Physical World." *Electronic Frontier Foundation*. https://www.eff.org/pages/leaving-physical-world.

Barlow, John Perry, and Mitch Kapor. 1991. "Civilizing the Electronic Frontier: An Interview with Mitch Kapor and John Barlow." Interview by David Gans and R.U. Sirius. *Mondo 2000*, Winter 1991, 45–50.

Benedikt, Michael. 1991. Introduction to *Cyberspace: First Steps*, edited by Michael Benedikt, 1–26. Cambridge, MA: MIT Press.

Broecker, William. 1975. "Climatic Change: Are We on the Brink of a Pronounced Global Warming?" *Science* 189, no. 4201: 460–63.

Carruth, Allison. 2016. "Ecological Media Studies and the Matter of Digital Technologies." *PMLA* 131, no. 2: 364–72.

Centre for Energy-Efficient Telecommunications (CEET). 2013. "The Power of the Wireless Cloud: An Analysis of the Impact on Energy Consumption of the Growing Popularity of Accessing Cloud Services via Wireless Devices." Univ. of Melbourne: CEET. https://ceet.unimelb.edu.au/publications /ceet-white-paper-wireless-cloud.pdf.

Chang, Alenda. 2019. *Playing Nature: Ecology and Video Games*. Minneapolis: Univ. of Minnesota Press.

Chun, Wendy Hui Kyong. 2006. *Control and Freedom: Power and Paranoia in the Age of Fiber Optics*. Cambridge, MA: MIT Press.

Chun, Wendy Hui Kyong. 2011. "Crisis, Crisis, Crisis, or: Sovereignty and Networks." *Theory, Culture, and Society* 28, no. 6: 91–112.

Deloria, Philip J. 1998. *Playing Indian*. New Haven, CT: Yale Univ. Press.

DeNardis, Laura. 2012. "Hidden Levers of Control: An Infrastructure-Based Theory of Internet Governance." *Information, Communication, and Society* 15, no. 5: 720–38.

DeNardis, Laura, and Francesca Musiani. 2016. "Governance by Infrastructure." In *The Turn to Infrastructure in Internet Governance*, edited by Francesca Musiani, Laura DeNardis, Derrick L. Cogburn, and Nanette S. Levinson, 3–24. London: Palgrave.

Diggins, Sarah, Hannah Welzbacker, and Claire Shinner. 2020. "How a Pandemic Made Rural Internet Even More Critical." *Indian Country Today*, May 23. https://indiancountrytoday.com/news/how-a-pandemic-made -rural-internet-even-more-critical-R5yATN9fn0mJU1o8mAaiFg.

Dyson, Esther, George Gilder, George Keyworth, and Alvin Toffler. 1994. "Cyberspace and the American Dream: A Magna Carta for the Knowledge Age." *Future Insight* 1, no. 2. Progress & Freedom Foundation. http:// www.pff.org/issues-pubs/futureinsights/fi1.2magnacarta.html.

Elmer-Dewitt, Philip, and David S. Jackson. 1994. "Battle for the Soul of the Internet." *Time*, July 25, 50. LexisNexis Academic.

Estus, Joaqlin. 2020. "Virus Widens Gap between Internet Haves, Have-Nots." *Indian Country Today*, May 3. https://indiancountrytoday.com/news /virus-widens-gap-between-internet-haves-have-nots-edAUSKUn8U-eGLVng 05BAA.

Farmer, Paul. 2006. *AIDS and Accusation: Haiti and the Geography of Blame*. Berkley: Univ. of California Press.

Gibson, William. 1984. *Neuromancer*. New York: Acc Books.

Gibson, William. 1987. *Count Zero*. New York: Ace Books.

Harvey, David. 2001. "Globalization and the Spatial Fix." *geographische revue* 3, no. 2: 23–30.

Hayles, N. Katherine. 1999. *How We Became Posthuman: Virtual Bodies in Cybernetics, Literature, and Informatics*. Chicago: Univ. of Chicago Press.

Heale, M. J. 2009. "Anatomy of a Scare: Yellow Peril Politics in America, 1980–1993." *Journal of American Studies* 43, no. 1: 19–48.

Hu, Tung-Hui. 2015. *A Prehistory of the Cloud*. Cambridge, MA: MIT Press.

Leary, Timothy. 1988. "The Cyber-Punk: The Individual as Reality Pilot." *Mississippi Review* 16, no 2/3: 252–65.

Lindskoog, Carl. 2018. "Perspective: How the Haitian Refugee Crisis Led to the Indefinite Detention of Immigrants." *Washington Post*, April 9. https://www.washingtonpost.com/news/made-by-history/wp/2018/04/09/how-the-haitian-refugee-crisis-led-to-the-indefinite-detention-of-immigrants/.

Mathews, Jay. 1982. "Economic Invasion by Japan Revives Worry about Racism." *Washington Post*, May 14. LexisNexis Academic.

Miller, Timothy. 1999. *The 60s Communes: Hippies and Beyond*. Syracuse, NY: Syracuse Univ. Press.

Moses, Peter, and John Moses. 1983. "Haiti and the Acquired Immune Deficiency Syndrome." *Annals of Internal Medicine* 99, no. 4: 565.

Native Business Staff. 2020. "Native Nations Own Broadband Spectrum on Tribal Lands: Haaland, Warren Bill to Grant Permanent Access." *Native Business*, July 28. https://www.nativebusinessmag.com/native-nations-own-broadband-spectrum-on-tribal-lands-haaland-warren-bill-to-grant-permanent-access/.

Reich, Charles A. 1970. *The Greening of America*. New York: Random House.

Roh, David S., Betsy Huang, and Greta A. Niu. 2015. "Technologizing Orientalism: An Introduction." In *Techno-Orientalism: Imagining Asian in Speculative Fiction*, edited by David S. Roh, Betsy Huang, and Greta A. Niu, 1–22. New Brunswick, NJ: Rutgers Univ. Press.

Sardar, Ziauddin. 1996. "alt.civilizations.faq: Cyberspace as the Darker Side of the West." In *Cyberfutures: Culture and Politics on the Information Superhighway*, edited by Ziauddin Sardar and Jerome R. Ravetz, 14–42. New York: New York Univ. Press.

Sohn, Stephen Hong. 2008. "Introduction: Alien/Asian: Imagining the Racialized Future." In "Alien/Asian," edited by Stephen Hong Sohn. Special issue, *Melus* 33, no. 4: 5–22.

Starosielski, Nicole. 2015. *The Undersea Network*. Durham, NC: Duke Univ. Press.

Stoll, Christian, Lena Klaaßen, and Ulrich Gallersdörfer. 2019. "The Carbon Footprint of Bitcoin." *Joule* 3, no. 7: 1647–61.

Tabacco, Christina. 2020. "Senator Introduces Native Broadband Spectrum Ownership Bill." *Law Street*, July 29. https://lawstreetmedia.com/tech/senator-introduces-native-broadband-spectrum-ownership-bill/.

Turner, Fred. 2006. *From Counterculture to Cyberculture: Stewart Brand, the Whole Earth Network, and the Rise of Digital Utopianism*. Chicago: Univ. of Chicago Press.

Walker, John. 1988. "Through the Looking Glass: Beyond 'User Interfaces'." In *The Art of Human-Computer Interfaces*, edited by Brenda Laurel, 439–48. Redding, MA: Addison Wesley.

Walser, Randal. 1991. "The Emerging Technology of Cyberspace." In *Virtual Reality: Theory, Practice and Promise*, edited by Sandra K. Hesel and Judith Paris Roth, 35–41. Westport, CT: Meckler Publishing.

Rebecca
Evans

Geomemory and Genre Friction:
Infrastructural Violence and Plantation
Afterlives in Contemporary African
American Novels

Abstract This essay argues that contemporary African American novels turn to the gothic in order to dramatize the uncanny infrastructural and spatial afterlives of the plantation through a literary strategy it identifies as *geomemory*: a *genre friction* between mimetic and gothic modes in which postplantation spaces in the US South are imbued with temporal slippages such that past and present meet through the built environment. Tracing the plantation's environmental and infrastructural presence in the Gulf Coast and throughout the US South, this essay argues that the plantation's presence is fundamentally gothic. Geomemory, a trope evident across the emerging canon of contemporary African American fiction, allows writers to address the representational challenge of infrastructural and spatial violence via a defamiliarizing chronotope in which past, present, and future come into uneasy contact. Further, geomemory's particular enmeshment with spatial design and infrastructure means that it moves from identifying the modern afterlife of the plantation to situating the present in the long context of plantation modernity.

Keywords African American literature, environmental humanities, gothic, Plantationocene

E
arly in Toni Morrison's novel *Beloved* (1987), Sethe formulates a theory of haunting. "Some things go. Pass on. Some things just stay. I used to think it was my rememory," she tells her daughter Denver. "But it's not. Places, places are still there. . . . Where I was before I came here, that place is real. It's never going away. Even if the whole farm—every tree and grass blade of it dies. The picture is still there and what's more, if you go there—you who never was there—if you go and stand in the place where it was, it will happen again; it will be there for you, waiting for you" (Morrison 2004: 43–44). Unlike the more familiar version of rememory, this version of supernatural and insistent historical return emanates not from human individuals but seemingly from the land itself—specifically, the land

American Literature, Volume 93, Number 3, September 2021
DOI 10.1215/00029831-9361265   © 2021 by Duke University Press

of the plantation.[1] This *spatialized*, rather than individualized, account of rememory thus becomes a way for Morrison (through Sethe) to name the "geographical imperative" that, Ruth Wilson Gilmore (2002: 16) argues, "lies at the heart of every struggle for social justice": "if justice is embodied, it is then therefore always spatial, which is to say, part of a process of making a place."

In this essay, I trace how spatialized rememory manifests as a contemporary trope in African American fiction that I call *geomemory*: the imbuing of postplantation space in the US South with gothic slippages in time (with past and present meeting through the built environment) and in agency (with land and infrastructure taking on uncanny powers and abilities to act). Taken together, these two features of geomemory become a way for writers to narratively concretize how the plantation haunts southern landscapes, how it continues to exert violent agency through infrastructure and design. In readings of two twenty-first-century novels by Black writers—Jesmyn Ward's *Sing, Unburied, Sing* (2017) and Colson Whitehead's *The Underground Railroad* (2016)—I show how Ward and Whitehead use the friction between mimetic and gothic modes to allow geomemory to enter literary narrative. More specifically, I argue that Ward's and Whitehead's gothic invocations of geomemory constitute a literary strategy for reckoning with the historically extensive relationship between the plantation and infrastructural violence. The haunted places in which geomemory operates in *Sing, Unburied, Sing* and *The Underground Railroad* do more than memorialize extraordinary sites of historical violence: they mark moments in which violence becomes infrastructural, part of environmental design itself. Geomemory doesn't just insist that historical violence continues to matter in the present; it insists further that such violence is fundamentally modern. It thus rewrites the plantation as a paradigmatic figure of US modernity, which continues to organize geography and infrastructure around and through anti-Black systems of value extraction with catastrophic ecopolitical results.

Gothic geomemory does several important things. Within these novels (as a chronotope), geomemory formally allows for literary engagement with the long postplantation emergency of anti-Black (and anti-Indigenous) violence: it represents structures and infrastructure in spectacular ways, thus acknowledging the experience of those who live in relation to necropolitical regimes while refusing to naturalize this violence or render it unremarkable. Within the sphere of literary studies (as a formal feature), geomemory names a major element of contemporary African American literature: it allows us to recognize the patterns of genre experimentation by which Black writers have turned to the gothic to craft counterhegemonic models of historical change and spatial (in)justice. Finally, within broader conversations around space, race, and infrastructure (as a concept), geomemory demonstrates how literary strategies can be leveraged to represent the eerie agency of infrastructure.

This last point leads me to the final critical terrain on which geomemory operates: geomemory is a literary strategy uniquely attuned to the logics of the Plantationocene. Proposed to identify the geological epoch in which humans have become global-scale forces of ecological destabilization—and proposed specifically as an alternative to "Anthropocene," which generally suggests a homogenizing species-wide scale and a particular focus on fossil fuels—*Plantationocene* identifies the violent and extractive capitalist systems that worked against Indigenous groups, as well as enslaved Black people, in the early modern world as the root of contemporary global environmental catastrophe.[2] To say that we live in the Plantationocene is to say that the crisis of climate change is rooted in the global system of enslaved plantation labor that structured European colonization of Africa and of the Americas; it is to say that we cannot understand our climate present without tracing how a residual logic of the plantation shapes contemporary extractive petrocapitalism; and it is to say that our climate crisis began both materially and ideologically with plantation modernity. By situating haunting in the spatial design of the plantation and by animating the plantation through contemporary infrastructures, geomemory becomes pedagogical, teaching readers how to read themselves into the Plantationocene. In this essay, I first trace how the plantation has haunted southern landscapes specifically through fossil fuel infrastructures. I then consider the historical role of the gothic in demanding acknowledgement of a form of violence that is not *invisible* but rather, per Katherine McKittrick (2011: 954), "longstanding but unacknowledged," before showing how gothic geomemory operates in Ward's and Whitehead's novels to render the realities of the Plantationocene.

Plantationocene Infrastructures

Even beyond the realm of the literary, it is far from metaphorical to suggest that former plantation spaces in the US South are haunted by their pasts. Indeed, southern ecological devastation is rooted not only in the intensive consumption of fossil fuels since the Great Acceleration (often dated to around 1950) but also in the legacy of the plantation system. This legacy is both ideological and material (McKittrick 2013: 3): major southern sites of environmental racism are often former plantation spaces with infrastructural "memories" that continue to perpetuate environmental racism. Willie Jamaal Wright's (2021: 793) description of "[r]acial landscapes produced by anti-Black violence" (804) is relevant here, as is his call for "a reconceptualisation of environmental racism that takes into account the mergence of anti-Black violence and ecologies," which is to say, a reframing of contemporary environmental racism in the broader and longer sweep of anti-Black spatial organization. Through this lens, grasping contemporary southern environmental racism requires not just an acknowledgment of relatively recent policy decisions leading to unevenly distributed environmental exposure, but also an acknowledgement of the long material histories of environmental and social disposability. In other words, the plantation is both so persistent and so insidious in southern spaces precisely because it has transformed to operate through energy and other infrastructures, propagating familiar forms of violence in the guise of new regimes.

As McKittrick argues, however, when approaching the plantation through critical geography, we must trace its continuities and continuances without collapsing all forms of spatial violence into a plantation monolith. She writes:

> I am not claiming that the plantation and contemporary geographies in the Americas are indistinguishable or identical. Rather I am positioning the plantation as a very meaningful geographic prototype that not only housed and normalized [. . .] racial violence in the Americas, but also naturalized a plantation logic that anticipated (but did not twin) the empirical decay and death of a very complex black sense of place. (McKittrick 2011: 951)

Infrastructure, I suggest, offers a useful heuristic to follow both the persistence and the mutation of the plantation in the US South. By considering the long life of the plantation in terms of infrastructure, we see how infrastructural projects, decisions, and designs have both perpetuated and naturalized the anti-Black and anti-Indigenous violence of the plantation throughout the twentieth century and into the twenty-first—how the plantation has continued to motivate and execute extractive logics and industries, with catastrophic global consequences. The infrastructural lens emphasizes that the plantation's contemporary agency is not simply a function of an extant but largely vanquished past violence; rather, the plantation was *always* a carefully designed spatial logic of modernity. As Sylvia Wynter (1971: 100–101) writes, "The plantation system, which, under the liberal Free Trade rhetoric, the rhetoric which freed the slaves, compensated the masters and set the slaves free in a world dominated by market relations, to fend naked for themselves, was the first sketch of monopoly capitalism." McKittrick (2011: 949) puts it even more starkly: "With a black sense of place in mind," she writes, "the plantation notably stands at the centre of modernity. It fostered complex black and non-black geographies in the Americas and provided the blueprint for future sites of racial entanglement."

Infrastructure, then, is one way to trace the plantation's continued agency as a blueprint for spatial violence and spatial relations—to consider the forced displacements, intensified extractions, and deliberately planned patterns of profitable sacrifice zones that characterize environmental racism. Infrastructure also serves as a useful heuristic because, like environmental racism itself, it is so often obscured within the hegemonic stories plantation modernity tells itself about itself: moments of spectacular failure aside, infrastructural access and infrastructural violence are rendered invisible, with racial violence naturalized and material causalities obscured (though never, of course, to those against whom such violence is perpetrated). In what follows, I explore two paradigmatic sites that reveal different ways in which the historical violence of the plantation has been (or, rather: has *always* been) built into environmentally catastrophic infrastructures.

Perhaps the better known of these sites is "Cancer Alley," the name given to the former plantation land, extending along the Mississippi River from New Orleans to Baton Rouge in Louisiana, that contains over one hundred high-emission petrochemical plants along its roughly eighty-five miles. Cancer Alley's plantation past shaped its petrochemical present: infrastructural and other features of plantation tracts (their proximity to the river; their convenient access to railways; their inhabitants' lack of economic and political capital) meant that Cancer Alley was for twentieth-century petrochemical plants an appealing site, able to be bought cheaply, developed profitably, and run with little oversight or regulation. These plants' toxic and carcinogenic outputs have produced increased respiratory complaints, prenatal issues including miscarriages, and clusters of cancer among its residents, many of whom are the descendants of enslaved people.[3] To those living in Cancer Alley, the continuity between historical plantation slavery and contemporary petrochemical exposure is clear. Sociologist Thom Davies (2017) writes, "The region's plantation past has been transposed onto the toxic geographies of today. As veteran environmental justice campaigner Darryl Maley-Wiley explained in an interview at the Sierra Club in New Orleans, the transfer from Plantation to Chemical Plant was effectively '*exchanging one plantation master for another.*'" Cancer Alley resident Amos Favorite expresses similar sentiments in a 1990 interview, saying, "We are all victimized by a system that puts dollars before everything else. That's the way it was in the old days when the dogs and whips were masters, and that's the way it is today when we got stuff in the water and air we can't even see that can kill us deader than we ever thought we could die" (Bullard 2018: 106). In their insistence that the petrochemical industry serves as a new kind of plantation master, community members work to draw attention to this historical continuity under racial capitalism.[4]

Cancer Alley reveals the direct infrastructural legacies of the plantation on former plantation grounds—but the plantation makes its infrastructural impacts known even in absentia. Take, for instance, the Isle de Jean Charles off the coast of Louisiana, the first recipient of federal funds intended for the relocation of an entire community due to impending climate displacement. The Isle de Jean Charles, currently home to about one hundred members of the Isle de Jean Charles Biloxi-Chitimacha-Choctaw Tribe (IDJC), is "sinking" due to sea level rise, shrinking from erosion, and vulnerable to storms; since 1955, the isle has lost 98 percent of its land. Following Neil Smith's (2006) famous formulation—"there's no such thing as a natural disaster"—and Kyle Powys Whyte's (2018) point that the prospect of ecological apocalypse looks quite different from an Indigenous perspective in which such an apocalypse has been unfolding for centuries, the residents of the isle have resisted all monolithic explanations of climate change that exclude more proximal human causes. Instead, the community-controlled website emphasizes infrastructural and social policies: "Our tribal lands," it reads, "are plagued by a host of environmental problems—coastal erosion and salt-water intrusion, caused by canals dredged through our surrounding marshland by oil and gas companies, land sinking due to a lack of soil renewal or 'crevasse,' because of the construction of levees that separated us from the river, and rising seas." Furthermore, as the site notes, the ancestors of the current residents were moved there as a result of "Indian Removal Act–era policies" (fewer than two hundred years ago), under which Indigenous people were forced to vacate space understood as agriculturally viable and move into "uninhabitable swampland."[5] This emptying of space in the South, of course, cleared room for the expansion of profitable plantation agriculture. The plantation thus doubly haunts the Isle de Jean Charles: first through the residents' history of forced displacement to land marked as less valuable; then through the plantation's petrochemical heirs and their infrastructural devastation of coastal lands, with canals and levees eroding and blocking the replenishment of the isle's soil.[6]

By reckoning with the infrastructural logics of both Cancer Alley and the Isle de Jean Charles, we can situate the contemporary crises of environmental racism in the US South in a longer account of Plantationocene modernity. This nomenclature (Plantationocene rather than Anthropocene) approaches climate change through the lens of Black critical geography to conclude, first, that responsibility for and vulnerability to environmental catastrophe is not evenly distributed across humanity and, second, that what appears to be a modern problem of energy technology is in fact a systemic one that stretches back to European colonization and slavery.[7] Both Cancer Alley and the Isle de Jean Charles show that the legacies of the Plantationocene cannot be reckoned with unless we grasp how the plantation system was originally rooted in spatial design and how, even as that system has transformed through varying regimes of racial violence, it has continued to operate through extractive infrastructures.[8]

The Plantationocene thus marks the uncanny temporality by which the plantation haunts structures of social relation, petrochemical legacies of monocrop extraction, forms of forced and unfree labor, and the organization and use of space.[9] Tracing these legacies is an infrastructural form of what, after Christina Sharpe (2016), we can call "wake work": an exploration of the extensive histories of enslavement in the Americas that have transformed but never ended, and that continue to drive global environmental catastrophe.

This durational violence is multidimensional: it is ideological, social, and political, yes, but it is also infrastructural and environmental. It is also temporally complex, both because of the philosophy of history it suggests and because infrastructure itself ports strange temporalities. As the introduction to the edited collection *The Promise of Infrastructure* (2018) by Nikhil Anand, Akhil Gupta, and Hannah Appel puts it, infrastructures are "spatiotemporal projects," "chronotopes"; they "configure time, enable certain kinds of social time while disabling others, and make some temporalities possible while foreclosing alternatives" (17, 15). Brian Larkin (2013: 333), for instance, reminds us that the global repetition of infrastructural projects allows different locales to "participate in a common visual and conceptual paradigm of what it means to be modern." Naming this hypervisible temporal model as "developmental time—linearity, progress, teleology," Hannah Appel (2018: 44) notes that infrastructure operates simultaneously in the quite different chronotope of "repetition and cyclicality; serial frontiers; abandonment, decommission, and ruins." As the resistant rhetorics of those living along Cancer Alley and on the Isle de Jean Charles demonstrate, naming the infrastructural violence of the plantation means wrestling with these chronotopes, bending time so that past and present come into uncomfortable contact and modernity is revealed as always embedded in historical violence. It is this defamiliarizing temporality and the uncanny agency of infrastructure that prompt contemporary literary writers who engage the plantation to turn to the gothic.

Gothic Geomemory

Just as Morrison uses gothic haunting to convey both the individual trauma that Sethe experiences and the spatial violence that haunts sites associated with slavery, so, too, have contemporary Black writers returned to the gothic to reveal the infrastructural and spatial roots of structural violence. The gothic imaginary of these contemporary novels, however, situates its uncanny animacy not on particular plantation sites but rather in particular forms of spatial, technological, and infrastructural design associated with plantation slavery. It is this phenomenon—infrastructural violence that appears to haunt sites of slavery in the US South, the interruption of otherwise mimetic texts by a fantastically ominous gothic animacy—that I call geomemory. Geomemory is especially evident in contemporary Black literature, where it takes various gothic forms: explicitly or implicitly cursed sites, ghosts and other uncanny spectral hauntings whose presence is narratively tied to specific places, and spatiotemporal oddities in which locations take on nonlinear or anachronistic temporal features.[10] Together, these manifestations coalesce around the uncanny distortion of spatiotemporal logic such that the temporal identity of a particular place *bends*. By integrating the gothic into realist fictions, contemporary Black novelists weave distributed understandings of historical causality, agency, and timescale into one story, and they use these hybrid stories to challenge the reductive progress narrative of liberal modernity (according to which regimes of racial violence have ended). These formal shifts between realism and the gothic, which I call *genre frictions*, are a literary strategy for integrating difficult knowledge into everyday life. Genre friction is one way in which the contemporary lit-erary field evinces "the changing status of genre fiction," as literary writers borrow from genres traditionally banished to the realm of "pulp" (Hoberek 2007: 240). As Jeremy Rosen (2018) argues, "writ-ers can modify the genres they utilize in pursuit of varied representa-tion goals"; thus, he suggests, it's crucial that critics avoid "attribut[ing] agency to the tropes of genre fiction" and to remember that "literary writers have been strategically deploying those tropes."[11] Taking seri-ously Rosen's point about authors' strategic genre decisions, this essay asks: toward what ends do contemporary Black writers deploy the gothic tropes of geomemory?

My choice of the label *gothic* speaks both to the genre's trans-historical thematic content—uncanny violence, historical resurgence, claustrophobia and entrapment—and to the gothic's historical enmesh-ment with representations of social and environmental horror. In the American tradition, the gothic has long been understood as engag-ing histories of racial violence, particularly white violence against Indigenous and enslaved Black people.[12] More recently, Sheri-Marie Harrison (2018) has proposed that a "New Black Gothic" is emerg-ing in works like Donald Glover's song "This Is America" (2018) and Jordan Peele's film *Get Out* (2017). These pieces, Harrison suggests, "[employ] Gothic tropes to embed contemporary developments such as mandatory minimum sentencing and the War on Drugs in a longer history of slavery and Jim Crow." This form of the gothic, however, does not end in exorcism or redemption through representation but instead "speak[s] to an ever-present and visible lineage of violence that accumulates rather than dissipates with the passage of time": "Gothic violence remains a part of everyday black life." Indeed, Harri-son suggests, there is something in the very tropes of the gothic—its insertion of brutality into experience, its insistence that, per southern gothic exemplar William Faulkner (2011: 73), "the past is never dead. It's not even past"—that is uniquely attuned to registering (though, crucially, never narratively resolving) the legacies of violence that have persisted through slavery and Jim Crow into neoliberal regimes of structural violence.[13]

Further, as a genre or mode, the gothic allows writers to bend the rules of the world in ways that are uniquely well suited for represent-ing environmental injustice. As Lawrence Buell (1998: 645) writes, the environmental imagination's contemporary shift from the pastoral to "toxic discourse"—which dates, he suggests, to Rachel Carson's description in 1962's *Silent Spring* of the effects of the indiscriminate use of chemicals—owes a significant debt to the gothic: "the more toxic discourse focuses on specific cases, the more readily toxic discourse montages into gothic" (653).[14] The gothic nature of environmental toxicity becomes even more pointed in the context of *unequal* exposure in the present based on historical violence in the past. Cancer Alley exemplifies uncanny animacy and historical resurgence, as nonhuman matter slowly but insistently enacts the racial violence that "haunts" the land alongside the Mississippi.[15] Under conditions of literary realism, landscapes don't and can't have *memory*; curses can't make violence emerge from nonhuman entities; the causal mechanisms of agency operate in linear and traceable ways.[16] Yet the gothic realities of geomemory are part of the reality of the Plantationocene.[17] As Elaine Gan et al. (2017: G2) ask in their introduction to *Arts of Living on a Damaged Planet*, "How can we get back to the pasts we need to see the present more clearly?" Their answer is profoundly gothic: for them, the figure of the landscape-specific "ghost" enables a "return to multiple pasts," a "haunting" that is "quite properly eerie." As they elaborate, "Ghosts show the layered temporalities of living and dying that shape our landscapes, tripping up the forward march of progress" (Swanson et al. 2017: M10–11); thus, "our monsters and ghosts help us notice landscapes of entanglement, bodies with other bodies, time with other times" (M7).

The particular phenomenon I call geomemory, meanwhile, builds on the gothic's acknowledgement of historical persistence by specifically rooting that persistence in the organization of space. Geomemory doesn't just show that history haunts us; it shows precisely how that haunting plays out in the human use and misuse of land and the human organization of space and infrastructure. The "geo" of geomemory is purposefully ambiguous, limning the geological and the geographical, because it is precisely this ambiguity that the Plantationocene emphasizes: as a concept, the Plantationocene reminds us that Anthropocene geology is the product of specific *geographies*; we cannot disentangle human geography, infrastructure, and spatial organization from the traces humans leave in the geological record. By imbuing space with uncanny temporality and unfamiliar agency, gothic geomemory teaches us to recognize the strange and estranging reality of infrastructural and spatial violence. Much as gothic violence in nineteenth-century Black literature strategically resituated horror in the real practice of enslavement rather than in imaginary moments of the supernatural, gothic geomemory in twenty-first-century African American novels dramatizes the interruptive nature of infrastructural violence, functioning almost pedagogically as it offers access to accurate but representationally challenging models of Plantationocene history.[18] In what follows, I show how the genre friction of gothic geomemory serves to illuminate the infrastructural presence of the plantation in Ward's *Sing, Unburied, Sing* and Whitehead's *The Underground Railroad*. In these novels, I argue, Ward and Whitehead harness the uncanny properties of the gothic in order to spectacularize the persistence of the plantation, creating a chronotope of plantation modernity via the strange agencies of infrastructure and spatial design.

Parchman's Plantationocene in *Sing, Unburied, Sing*

Sing, Unburied, Sing follows teenage Jojo and his younger sister Kayla as they embark on a journey across Mississippi with their mother, Leonie. Leonie, who leaves the children mostly on her parents' farm, is taking them to pick up their father, Michael, a white man, as he is being released from the prison known as Parchman Farm, now officially named Mississippi State Penitentiary. Much of the text is realist in scope, focusing on the experience of the journey to and from Parchman itself: we learn of Kayla's stomach ailments, Leonie's efforts to transport drugs while avoiding police detection, and the passengers' strained relationships—both to each other, as Michael reunites with his partner and his somewhat skeptical children, and to the people they visit along the way, including Michael's racist parents, who prompt the family's exit when they target the children with a vicious outburst of white supremacy.

However, weaving in and out of this family-focused realism is a canonical gothic element: haunting. Two ghosts visit the competing narrators of the novel. Leonie is haunted by her brother, Given, who was killed years before at the age of eighteen by one of Michael's racist cousins in what was ostensibly an accident on a hunting trip. Jojo's ghost is also a murdered young Black man, but JoJo never knew his ghost in life and doesn't learn his specter's history until later in the book. This latter ghost is Richie, who was imprisoned as a preteen at Parchman with Leonie's father Pop (Jojo's grandfather and essentially his acting father) and died there. Richie's death, we learn, occurred in the course of an unsuccessful prison escape attempt; much later in the book, we are given the full story, which Pop has held back for years: Richie was running away with a man who had committed sexual assault, and Pop, who had been tasked with overseeing the dogs, was charged with tracking them down. Pop found Richie after his fellow fugitive had been captured and tortured to death by a vicious lynch mob; Richie died by Pop's hand, a mercy killing that came unexpectedly and saved the young man from a much more prolonged and painful end.

These two ghosts echo the divergent explanations of haunting generated within *Beloved*. Given is linked specifically to Leonie's own memories and experience, and he follows *her*; he embodies direct violence, the explanatory model of specific ghosts manifesting Sethe's personal trauma and guilt. By contrast, Richie is tied to Jojo not because of Jojo's personal experiences but rather by a complex legacy of familial and spatial trauma. He comes to Jojo because of their shared connection to Pop, but he is able to converse and travel with Jojo only after Jojo is brought to Parchman, where both his white father and his Black grandfather have been incarcerated, and where Richie died. Richie, then, is both the ghost of an individual who experienced violence *and* the ghost of the spatial legacies held within the site of Parchman, and the shifting regimes of violence that the prison represents.

The temporal slippage of Richie's ghostly life lets him teach the reader to locate the present injustices of Parchman within much longer histories. We see this in a chapter he narrates:

> I despaired, burrowed into the dirt, slept, and rose to witness the newborn Parchman: I watched chained men clear the land and lay the first logs for the first barracks for gunmen and trusty shooters. I thought I was in a bad dream. I thought that if I burrowed and slept and woke again, I would be back in the new Parchman, but instead, when I slept and woke, I was in the Delta before the prison, and Native men were ranging over that rich earth, hunting and taking breaks to play stickball and smoke. Bewildered, I burrowed and slept and woke to the new Parchman again, to men who wore their hair long and braided to their scalps, who sat for hours in small, windowless rooms staring at big black boxes that streamed dreams. Their faces in the blue light were stiff as corpses. I burrowed and slept and woke many times before I realized this was the nature of time. (Ward 2017b: 186–87)

Richie's vision of the site that binds him, and by extension his very presence in *Sing, Unburied, Sing*, brings to Jojo and to the reader a long history of Parchman. As he slips through history, Richie's spectral reverie identifies Indigenous dispossession, post-Emancipation Black convict labor, and the contemporary carceral state as part of the same historical system—and as inexorably tied to the sites on which that system's violence manifested. Time travel—both backward into the past before his own life, and forward as a ghost into the post–Jim Crow future of American incarceration—lets Richie embody, and teach the reader to inhabit, a form of history guided by *continuity* rather than progress. "Sometimes I think it [Parchman] done changed," he tells Jojo. "And then I sleep and wake up, and it ain't changed none . . . It's like a snake that sheds its skin. The outside look different when the scales change, but the inside always the same" (172). Whenever he awakens in time, what Richie sees is inevitably all part of the same historically extensive trauma that Parchman contains, which conveys how the violence that stretches from plantation to Parchman is bound up in spatial organization, land usage, and environmental infrastructure.

Parchman was constructed in the early years of the twentieth century but, as historian David M. Oshinsky (1996: 137) writes, "looked like a typical Delta plantation, with cattle barns, vegetable gardens, mules dotting the landscape, and cotton rows stretching for miles"; thus, "In design, it resembled an antebellum plantation with convicts in place of slaves" (139). In a National Public Radio (NPR) interview, Ward (2017a) cited Oshinky's book on Parchman as inspiration for the character of Richie, recalling her realization that "I have to write about a kid like this, and this child has to have agency [. . .] in the present moment" as the moment when she "understood that I was writing a ghost story too." In Ward's own account, her novel's gothic element of haunting was a direct response to historical knowledge about the horrific violence of Parchman, a literary strategy for representing history without replicating historical erasure.

The relationship between Parchman and plantation slavery was more than visual: originally a segregated prison built to replace the convict leasing system, its construction dovetailed with southern policies of racist incarceration that almost seamlessly transformed plantation slavery into prison slavery. Richie, Pop, and other Black men in Parchman were jailed arbitrarily for petty or invented crimes and put to work toiling in fields making cash crops; their labor was forced and unpaid, their working and nonworking hours supervised under the threat of constant violence, and their deaths threatened and meted out without consequence. Their incarceration, the novel makes clear, was a direct extension of slavery: as Ward (2017a) said in the same NPR interview, "Parchman prison was basically a big plantation in the 1930s, the 1940s."

The challenge is representing not only these *histories* but also how they continue to structure violence in the present. Like Cancer Alley and the Isle de Jean Charles, Parchman reveals how the plantation haunts the landscape, how it surfaces in persistent infrastructure, in post-Emancipation design, and in the disproportionate vulnerability of Black and Indigenous people to environmental hazards—hazards that, as David Naguib Pellow (2018) documents, are often forced on incarcerated people through infrastructural exposures such as water and air contamination, dangerous heat conditions, toxic on-site industrial labor programs such as e-waste recycling, and proximity to military and other waste dumps. Parchman, then, contains both the legal/juridical/ideological and the spatial/infrastructural/environmental continuity between plantation slavery and the carceral state. The ways in which plantation slavery transformed into prison labor have become part of the cultural lexicon through the critical and popular success of works like Michelle Alexander's *The New Jim Crow: Mass Incarceration in the Age of Colorblindness* (2010) and Ava DuVernay's documentary film *13th* (2016). Yet as the history of Parchman reminds us, the racial violence of contemporary incarceration is bound to the plantation through design as well.

In giving us a gothic version of Parchman via a ghostly narrator, Ward thus uses the uncanny historical slipperiness of haunting to situate contemporary incarceration in longer spatial and environmental histories of violence. This is rememory as infrastructural violence, haunting as geomemory, and gothic as a form of historical imagination that refuses presentism and situates contemporary injustice in long American legacies of the violence of the plantation. "How could I know," Richie asks, "that after I died, Parchman would pull me from the sky? How could I imagine Parchman would pull me to it and refuse to let go? And how could I conceive that Parchman was past, present, and future all at once? That the history and sentiment that carved the place out of the wilderness would show me that time is a vast ocean, and that everything is happening at once?" (186). Here, Richie vocalizes the metaquestion that the novel itself is asking: how can we *know*, how can we *imagine*, and how can we *conceive* of the infrastructural legacies of the plantation? Ward's novel shows us that we need the gothic in order to grapple with structures of violence that shape the US South: the gothic allows us to inhabit a form of historical imagination in which we can see the continuities between past and present, beyond the narrower scale of individual experience.

Sing, Unburied, Sing is not primarily concerned with energy infrastructure. The closest it comes is attributing Michael's incarceration indirectly to his experience of the Deepwater Horizon oil spill of 2010, after which he sank into addiction, and allowing Michael's discussion of that event with his son to linger on the corporate scientific denial of the multispecies fatalities of that spill, with the suggestion that this denial taught Michael about the differential disposability of certain lives, human and nonhuman, under industrial modernity (226).[19] Nonetheless, Deepwater's appearance in this novel underscores the infrastructural logics of continuity that motivate Ward's use of geomemory. The event figures not only the ecological and human catastrophe that Michael witnesses but also the larger systems of extraction and disposability that the Plantationocene names. Michael's own revelatory experience with petro-infrastructure through Deepwater Horizon articulates the connections from the plantation to the petroleum complex as well as to the prison. The infrastructural trauma that brought Michael to Parchman ties the environmental injustices of Cancer Alley to the globally scaled slow violence of climate change. Deepwater's appearance in a novel dedicated to plantation infrastructure underscores, then, the long history of plantation modernity and the challenges of accurately capturing the Plantationocene.

Sing, Unburied, Sing thus uses the genre friction between gothic and realist modes to curate a readerly experience of the ways in which specific sites and forms of design build violence into the organization of space, with consequences that resonate from the regional to the global. The novel offers a model of how literary geomemory can teach readers to reckon with the difficult knowledge of infrastructure, to contend with how built environments render the complex agential forces of structural and infrastructural violence "natural" and unremarkable. Dwelling within a temporal structure of haunted simultaneity, Richie teaches the readers to see how Parchman's present regimes of violence are properly situated in the various extractive logics that have physically and socially shaped the Gulf Coast, from the time when violent sites were "carved . . . out of the wilderness" to the present.

Excavating the Present in *The Underground Railroad*

In contrast to the canonical haunting of *Sing, Unburied, Sing*, in which ghosts interrupt the (narrative and actual) present, in Whitehead's *The Underground Railroad*, geomemory's distortion manifests as the future (the reader's present) disruptively emerging into the narrative present (the reader's past). These spatiotemporal distortions produce a gothic model, not of the *presence* of the past (which is to say: the afterlife of the plantation) but rather of the *present-ness* of the past (which is to say: the present's proper location within a long history of plantation modernity).

A work of historical fiction, albeit one with (as I will discuss) an unusual relationship to historical fidelity and periodization, *The Underground Railroad* follows Cora, who was born into slavery on the Randall plantation in Georgia. Cora's mother, Mabel, escaped when Cora was a child and was never caught; Mabel's abandonment of her daughter haunts Cora, while the mysterious nature of Mabel's apparently successful fugitivity continues to gall the vicious Randall brothers and Ridgeway, a slave catcher and the horror-movie villain of the novel. The book's first section narrates Cora's various acts of resistance at Randall, from revenging herself against the enslaved man who casually destroys the provision ground that is her family legacy, to intervening in the beating of a young boy, to joining another enslaved man, Caesar, on a quest for freedom. While the third member of their party is captured and returned to Randall, Cora and Caesar continue their journey to the underground railroad stop about which Caesar has been told. The railroad ferries them to South Carolina, where they live in relative peace for a time before Caesar is captured and (we later learn) killed. Cora, however, manages to pick up the railroad again, and in North Carolina, she is taken in by an aging couple who conceal her in a nightmarishly claustrophobic attic crawl space lifted from the pages of Harriet Jacobs's 1861 *Incidents in the Life of a Slave Girl*. From this prison, Cora silently observes the theatrical genocide by which North Carolina perpetuates its all-white state; Irish immigrant laborers carry out the work formerly done by enslaved Black people while the latter (and all those who harbor them) are murdered in grotesque and ritualistic performances. Eventually, Ridgeway catches up with Cora. He gives her former protectors over to the town's whims and brings Cora on a side trip to catch a fugitive in Tennessee before returning her to Randall. In Tennessee, Cora is rescued by a group of free Black people who take her with them along the underground railroad to the Valentine farm in Indiana, a utopian community of free and escaped Black people. However, Ridgeway finds her again, and Cora watches the destruction of the Valentine farm before finally managing to kill Ridgeway, escape along the last leg of the underground railroad, and emerge. The last scene sees her joining a group of pioneers heading to California via Missouri.

The uncanny temporal oddities of geomemory bring the twentieth century into the nineteenth in two ways in Whitehead's novel. First, each of the states to which Cora travels encodes different regimes of racial violence; wherever Cora stops, she encounters strange worlds that twenty-first-century readers can recognize as integrating elements of postslavery white supremacy. "Every state is different," the first underground railroad operator Cora encounters tells her and Caesar: "Each one a state of possibility, with its own customs and ways of doing things. Moving through them, you'll see the breadth of the country before you reach your final stop" (Whitehead 2016: 68–69). In South Carolina, for instance, Cora encounters an apparently benevolent form of governmental control over Black labor, wherein the government owns all Black residents yet allows them to live independent lives. Yet the peaceful façade hides new forms of racial violence that twenty-first-century readers recognize as references to medical experimentation on Black people that continued through the twentieth century into at least the 1970s. As Cora learns, some of the "many studies and experiments under way at the colored wing of the hospital" include anachronisms such as the forced sterilization of Black women deemed "unfit" (which continued through the 1970s) and the Tuskegee Study of Untreated Syphilis in the Negro Male (which ran from the 1930s through the 1970s).[20] Doctors describe these programs as triumphs of modernity, "one of the boldest scientific enterprises in history": "was it any wonder the best medical talents in the country were flocking to South Carolina?" (121–22). Here, as in the South Carolinian twelve-story skyscraper equipped with an elevator, as in the similarities Julian Lucas (2016) traces in the *New York Review of Books* between the lynchings Cora observes in North Carolina and the 1898 Wilmington Massacre, time bends such that the present is anticipated in the past: the twentieth and twenty-first centuries are refigured not as the triumph of modern progress over the barbarity of slavery but instead as the half-concealed continuation of the "peculiar institution." Indeed, Cora's own introspection underscores the anticipation of modern regimes of racism by the plantation. When she first arrives in South Carolina, "[s]he put the plantation behind her," (96) but as she realizes, the plantation "haunted" (105) her: "[s]he did not live there anymore" (96) but "[i]t lived in them" (106).[21] By tying these temporal slippages to Cora's spatial movements, Whitehead constructs an uncanny chronotope in which southern space not only holds haunted *memories* of violence but also *anticipates* the illusory transformations of genocidal white supremacy into the seemingly more anodyne shapes it took in the twentieth century. Embedding nonlinear allusions to post-Emancipation moments of white violence in each of the places to which Cora travels, Whitehead imbues place with a form of memory that points forward. The strange slippages of geomemory show how plantation logics persisted into the spatially organized and institutionalized terrorization of Black people in the twentieth century; further, they show that this persistence was never an antiquated relic of the antebellum past, but rather remains the self-appointed center of American modernity.

Second, in the novel's titular and most spectacularly anachronistic conceit—the literalization of the underground railroad as a sub-terranean railway system—these geomemorial slippages in time, these uncanny gothic accounts of plantation *modernity*, are tied explicitly to infrastructure. We see this from the first moment Cora and Caesar real-ize the nature of the underground railroad and stare in wonderment:

> The stairs led onto a small platform. The black mouths of the gigan-tic tunnel opened at either end. . . . The sheer industry that had made such a project possible. Cora and Caesar noticed the rails. Two steel rails ran the visible length of the tunnel, pinned into the dirt by wooden crossties. The steel ran south and north presumably, springing from some inconceivable source and shooting toward a miraculous terminus. (67)

"Who built it?" they ask; "Who builds anything in this country?" the stationmaster replies (67). In this revelatory scene, as elsewhere, Cora's introspection prods the reader toward the crucial line of thought:

> Cora couldn't pay attention. The tunnel pulled at her. How many hands had it required to make this place? And the tunnels beyond, wherever and how far they led? She thought of the picking, how it raced down the furrows at harvest, the African bodies working as one, as fast as their strength permitted. The vast fields burst with hundreds of thousands of white bolls, strung like stars in the sky on the clearest of clear nights. When the slaves finished, they had stripped the fields of their color. It was a magnificent operation, from seed to bale, but not one of them could be prideful of their labor. It had been stolen from them. Bled from them. The tunnel, the tracks, the desperate souls who found salvation in the coordination of its stations and timetables—this was a marvel to be proud of. She wondered if those who had built this thing had received their proper reward. (68)

Imaginatively linking the literal infrastructure of the railroad with the extractive logics of the plantation world above, Cora sets up a juxtaposition that she will return to throughout the novel: the cognitively estranging marvel of the infrastructure of the underground railroad, built by Black people for the project of Black liberation, against the extractive systems that carry out the violent logics of the plantation. The speculative estrangement of the underground railroad thus provides the foundation of an extended infrastructural metaphor; the novelty of the railroad lets Whitehead lay bare the historical continuity of racist American regimes of energy, value generation, and exploitation and the ways in which the apparent modernity of those regimes, and of their environments and infrastructures, belies their plantation origins. Genre friction between realist and speculative modes allows the gothic science fictionality of both historical bending (described above) and uncanny infrastructure to become clear.

Cora takes us from Indigenous dispossession ("The land she tilled and worked had been Indian land. She knew the white men bragged about the efficiency of the massacres, where they killed women and babies, and strangled their futures in the crib") to pre-Emancipation plantation slavery ("Stolen bodies working stolen land. It was an engine that did not stop, its hungry boiler fed with blood") to post-Emancipation biopolitics ("With the surgeries that Dr. Stevens described, Cora thought, the whites had begun stealing futures in earnest") (117). Her rendering of the explanation her North Carolinian protector offers of that state's all-white regime echoes this technological and infrastructural metaphor:

> As with everything in the south, it started with cotton. The ruthless engine of cotton required its fuel of African bodies. Crisscrossing the ocean, ships brought bodies to work the land and to breed more bodies.
>
> The pistons of this engine moved without relent. More slaves led to more cotton, which led to more money to buy more land to farm more cotton. (161)

Even after North Carolina's shift from Black to Irish laborers, Cora knows, the plantation's logics had persisted; and even once the Irish achieved the fictionalized status of whiteness, thus escaping plantation labor, "a new wave of immigrants would replace the Irish, fleeing a different but no less abject country, the process starting anew. The engine huffed and groaned and kept running. They had merely switched the fuel that moved the pistons" (171).

This infrastructural and mechanical sense of energy and value generation is not a fantasy limited to Cora: even Ridgeway, the embodiment of slavery, imaginatively represents the modernity of the plantation through the metaphor of infrastructure. His blacksmith father is susceptible to a spiritual account of his profession as "rendering order," a nobler pursuit from whose enmeshment in plantation slavery he seems to shy (80): "Liquid fire was the very blood of the earth. It was his mission to upset, mash, and draw out the metal into the useful things that made society operate: nails, horseshoes, plows, knives, guns. Chains. Working the spirit, he called it" (73). Shedding his father's sentimentalism about grander professional meaning, Ridgeway sees even more clearly the ways in which the infrastructures of extractive modernity and the enslaved people laboring in the plantation system cannot be disentangled. "One [his blacksmith father] made tools," he thinks, "the other [Ridgeway himself] retrieved them": "The crop birthed communities, requiring nails and braces for houses, the tools to build the houses, roads to connect them, and more iron to keep it all running. Let his father keep his disdain and his spirit, too. The two men were parts of the same system, serving a nation rising to its destiny" (76). The infrastructural conceit of the underground railroad throws into relief the infrastructural reality of racial violence, the inextricability of modern technoculture and modern infrastructure from both the logics and the material structures of plantation slavery. It is fitting, then, that Cora's acts of violence against Ridgeway are all tied to the infrastructure of Black ingenuity and resistance. When, in Tennessee, she uses her chains to strangle Ridgeway until he is incapacitated, "Her scream came from deep inside her, a train whistle echoing in a tunnel" (226). And when she finally vanquishes Ridgeway in Indiana, in the fiery ashes of the Valentine farm, she does so by weaponizing the railroad itself, entwining herself with him and hurling both of them down the steep stairs into the tunnel that obsesses him as an emblem of the fugitivity he has made it his mission to ensnare.

Coda: Geomemory, Ecological Catastrophe, and Historical Imagination

In Whitehead's version of geomemory, the direction of temporal distortion may be reversed from the gothic haunting of Ward's novel, but the outcome is the same: an uneasy hinging of past and present, mobilized by the literary emergence of spectacular genres tied specifically to the geographies of slavery, such that the plantation's material, spatial, and infrastructural logics cannot be extricated from contemporary American life. These novels thus deploy the gothic in order to narrate the defamiliarizing chronotope of plantation modernity and to tie such modernity specifically to infrastructural technology and spatial design. Gothic geomemory allows these novels to literalize how the plantation has haunted US landscapes across centuries, structuring the spatial distribution of environmental and social vulnerability, infiltrating both the cultural and economic logics and the material infrastructures of extraction and disposability.

Gothic geomemory thus serves as a formal literary strategy for spectacularizing the forms of violence that plantation modernity seeks to naturalize, for puncturing a progressive model of temporality with the insistent interruption of the present by the past and the lurking presence of the present in the past. Through this disruptive bending of time, which centers Black futurity as much as anti-Black history, geomemory avoids the naturalization of violence that Katherine McKittrick (2013: 10) criticizes in much plantation discourse and instead recognizes the plantation for what it is: "a persistent but ugly blueprint of our present spatial organization that holds in it a new future." Even when not engaging infrastructure directly, then, contemporary Black novelists who use the gothic trope I call geomemory construct a historical imagination capable of not merely reproducing but of representationally *contesting* the challenging realities of spatial and infrastructural violence in the Plantationocene.

The temporal and agential distortions of gothic geomemory reframe contemporary forms of structural and infrastructural violence as endemic to the plantation, putting into narrative what the term Plantationocene puts into nomenclature: that the global petrocrisis of the present cannot be thought without engaging the infrastructural logics of the plantation past. In animating Plantationocene infrastructure, geomemory reminds us that the emergency in which we live has been a long one indeed and that not only the extractive logics but also the extractive infrastructures of fossil fuel modernity were born of the plantation. The point of geomemory is not simply that the past is *still* present; rather, it is that historical regimes of violence have always been modern. Bending both time and agency, geomemory charts through literary narrative an infrastructural imagination capable of reckoning with the present, not as a new catastrophe but instead as an ecopolitical catastrophe coterminous with Western modernity itself.

Rebecca Evans is an assistant professor of English at Southwestern University. Her current book project, "Structural Frictions: Environmental Violence and Contemporary American Narratives," explores the formal and political dimensions of genre hybridity in American literature, identifying structural violence as both a defining feature of contemporary life and a driving force behind contemporary genre formation.

Notes

1 Much has been written about how rememory represents melancholic forms of both individual trauma and historical violence that refuse to stay safely quarantined in the distant past. Avery F. Gordon's *Ghostly Matters* (1997) offers the definitive version of this argumentative line. As Gordon (2008: 139) writes, *Beloved* is "one of the most significant contributions to the understanding of haunting," theorizing "haunting and . . . the crucial way in which it mediates between institution and person" (142). See also Brogan 1998.

2 See Haraway 2015 and Haraway et al. 2016. Since Donna Haraway's discussion of the term, it has gained critical popularity, as in the University of Wisconsin's 2019–2020 Sawyer Seminar "Interrogating the Plantationocene."

3 For an account of the history of community organizing and resistance to environmental racism in the region, see Wright, Bryant, and Bullard 1997.

4 Examples of this direct identification of contemporary environmental exposure with legacies of plantation slavery abound: see the United Church of Christ Commission for Racial Justice's (1998) report on St. James Parish, Louisiana, titled *From Plantations to Plants*; Robert D. Bullard's (2018: 108–9) reflection in *Dumping in Dixie* on the plantation house visible on the Dow Chemical plant; and Richard Misrach's (2012) photograph *Ashland-Belle Helene Plantation, Acquired by Shell Chemical*. These varied texts trace how the necropolitical logics of the plantation system still shape the infrastructural identity of Cancer Alley.

5 *Isle de Jean Charles*. "The Island." www.isledejeancharles.com/island (accessed April 9, 2021).

6 Indeed, in January 2019, the IDJC refused the federal funds used to purchase land for the relocation, pointing out that the land had been selected and purchased without input from tribal leadership and that the relocation terms involved troubling denials of access rights to the isle to residents once they moved. This refusal spoke explicitly to the persistence of spatial histories of plantation-driven displacement, denying the common and erroneous popular portrayal of the resettlement project as an exclusively climatological phenomenon to which the United States government had responded with innovative and just measures.

7 In *A Billion Black Anthropocenes or None* (2018), Kathryn Yusoff connects these two critiques, following what she argues is a foundational geological violence from the false universalism of the naming of the Anthropocene epoch back to the geological imperatives of colonization and slavery in the Atlantic world.

8 As Natalie Aikens, Amy Clukey, Amy K. King, and Isadora Wagner (2019) argue, "petrochemical plants, military bases, car manufacturers, oil rigs, fish canning operations, nuclear testing sites, mineral extraction mines, [and] cloud data server warehouses" operate upon the same grounds as plantations, use the same "monocultural tradition of extractive logics," and exploit the labor of the descendants of enslaved people.

9   McKittrick (2013: 9) notes the way that the plantation bends time: "In many senses the plantation maps specific black geographies as identifiably violent and impoverished, consequently normalizing the uneven production of space. This normalization can unfold in the present, with blackness and geography and the past and the present enmeshing to uncover contemporary sites of uninhabitablity. Yet to return to the plantation, in the present, can potentially invite unsettling and contradictory analyses wherein: the sociospatial workings of antiblack violence wholly define black history; this past is rendered over and done with, and the plantation is cast as a 'backward' institution that we have left behind; the plantation moves through time, a cloaked anachronism, that calls forth the prison, the city, and so forth."

10   In the larger project from which this essay is drawn, I also consider geomemory's appearance in such works as Junot Díaz's (2007: 1) novel *The Brief Wondrous Life of Oscar Wao*, where sugarcane plantations in the Dominican Republic become a site through which the uncanny agency of what Díaz calls *fukú americanus* ("the Curse and the Doom of the New World") enacts its violence.

11   Rosen's primary goal is to disaggregate the strategic use of genre frameworks in critically acclaimed literary fiction from the status of genre fiction in general; he calls attention to the ways in which writers "avail themselves of the generative capacity and plasticity of genre frameworks such as SF, fantasy, detective, and zombie novels" yet "also take pains to mark their literariness by deploying recognizable literary techniques and by differentiating themselves from, often by denigrating, the lion's share of popular culture's voluminous output." In my own approach to the strategic genre flexibility of contemporary literary fiction, I am less interested in assessing whether or to what extent writers have transgressed or remained within existing literary niche markets than I am in exploring precisely what the "generative capacity" of speculative genres affords these writers in the context of social and environmental injustice.

12   As Leslie A. Fiedler (1997: 29) claims in *Love and Death in the American Novel* (1960), American fiction is "bewilderingly and embarrassingly, a gothic fiction, nonrealistic and negative, sadist and melodramatic—a literature of darkness and the grotesque in a land of light and affirmation." Likewise, Morrison (1992: 36) asserts in *Playing in the Dark* that the "gothic romance" allowed for "the head-on encounter with very real, pressing historical forces and the contradictions inherent in them." Meanwhile, as Teresa A. Goddu (1997: 8) characterizes the genre, the gothic represents "the hauntings of history," depicting "the historical horrors that make national identity possible yet must be repressed in order to sustain it" (10).

13   There is, as this account suggests, a good deal of overlap between twenty-first-century gothic geomemory and the earlier twentieth-century US tradition of southern gothic, with Morrison's *Beloved* standing, perhaps, at the midpoint of these two movements. Both share a preoccupation

with violence, especially racial violence, as well as an attention to the persistence of history against teleological accounts of modern progress; further, both work specifically to remap spatial imaginaries so as to recognize a historically deep sense of southern place. As my subsequent readings shall make clear, however, there are a number of formal, contextual, and thematic factors that lead me to see gothic geomemory as distinct from southern gothic. One is the more varied temporal oddities that characterize the chronotope of gothic geomemory in contemporary Black fiction: here, time bends in many directions, with the present haunting the past as much as the past haunts the present. Another is the conscious remapping of the "South": varied instances of gothic geomemory not only situate the US South in relation to the global South but also consciously trace pre- and post-Emancipation racial violence through the US North rather than focusing primarily on former Confederate states. Finally, while in southern gothic writing the gothic is a mood and a mode that permeates throughout, gothic geomemory tends to be unevenly and strategically distributed throughout otherwise realistic texts. It is this uneven distribution that informs my naming of this phenomenon as genre *friction* and my characterization of it as a strategy that writers adopt in order to invoke and to subvert readers' expectations about genre.

14   Even beyond this more specific documentary gothicism of toxic discourse, the fable that opens *Silent Spring* evinces the gothic: "Then a strange blight crept over the area and everything began to change. Some evil spell had settled on the community: mysterious maladies swept the flocks of chickens; the cattle and sheep sickened and died. Everywhere was a shadow of death" (Carson 2002: 2).

15   Particularly in its more supernatural forms, the gothic can also be understood as the literary form of new materialism or object-oriented ontology: its uncanny animations make room in fiction for what Jane Bennett (2010) calls "vibrant matter." I point out this parallel not to center new materialism but to suggest that it is much less new than its name implies. Uncovering the literary tradition of the environmental gothic also highlights rich archives of material and embodied forms of understanding, often produced by women writers and writers of color who were interested neither in reinforcing bounded models of humanist subjectivity nor in offering a depoliticizing and ahistorical posthuman redistribution of agency, but rather in insisting upon the centrality of material analysis and experience. As Kyla Wazana Tompkins (2016) suggests, the alleged novelty of new materialism is suspect: Black feminist and postcolonial criticism also offer material analyses and critiques of Western humanism, but unlike much of new materialism, these accounts do not obscure questions and histories of power and identity.

16   This is not to say that realism, with the range of subgenres and movements that fall within its rough contours, is not invested in tracing the complex causality of social injustice. Indeed, from the late nineteenth century through at least the middle of the twentieth, American naturalist

writers took a particular interest in representing the workings of social determinism, including the unevenly distributed exposure of environmental risk and vulnerability. Naturalism, though, tends toward a presentist descriptive focus, showing how durational and systemic social forces play out in a relatively short span of time: its scale is typically within the span of a single human life and is often much more temporally constrained. Thus, while realist and naturalist texts often engage structural violence in the present—indeed, as I discuss later in this essay, realist and naturalist modes do precisely that in Ward's and Whitehead's novels—I am particularly focused on addressing how the gothic bends time in order to incorporate more expansive models of plantation modernity than the presentist scale of realism and naturalism tends to allow.

17 Recent years have seen a growing interest in the intersection of ecocritical and gothic discourse, often called the *ecogothic* (see Estok 2009; Sivils 2013; Smith and Hughes 2013; Del Principe 2014; and Keetley and Sivils 2017).

18 See chapter 6 of Goddu's *Gothic America* (1997), for instance, for an account of how slave narratives including Harriet Jacobs's *Incidents in the Life of a Slave Girl* (1861) and Frederick Douglass's *Narrative of the Life of Frederick Douglass* (1845) invoke gothic horror in the context of white violence against enslaved people, using the gothic to indicate the unspeakable and unimaginable nature of this violence without allowing it to remain unspoken or unimagined.

19 I am grateful to Martyn Bone (2020) for making this point in a paper he delivered at the Modern Language Association convention in Seattle during the *Plantationocene versus Anthropocene: Global South Perspectives* session in which both of us presented.

20 See Harriet A. Washington's *Medical Apartheid: The Dark History of Medical Experimentation on Black Americans from Colonial Times to the Present* (2006), in particular its accounts of the eugenics movement in the early twentieth century as it targeted Black women for sterilization, often without their knowledge, let alone their consent, and of James Marion Sims, who invented major modern gynecological instruments and procedures by experimenting without anesthesia or analgesic on enslaved Black women.

21 As she realizes, the Black women "believing themselves free from white people's control and commands about what they should do and be . . . were still being herded and domesticated. Not pure merchandise as formerly but livestock: bred, neutered. Penned in dormitories that were like coops or hutches" (124). This activation of Cora's political consciousness follows directly on another moment in which she recognizes the truth of temporal continuity over modern progress: "Cora thought back to the night she and Caesar had decided to stay, the screaming woman who had wandered into the green when the social came to the end. 'They're taking away my babies.' The woman wasn't lamenting an old plantation injustice but a crime perpetrated here in South Carolina. The doctors were stealing her babies from her, not her former masters" (123).

References

Aikens, Natalie, Amy Clukey, Amy K. King, and Isadora Wagner. 2019. "South to the Plantationocene." *ASAP/J: The Open-Access Platform of ASAP/Journal*, October 17. http://asapjournal.com/south-to-the-plantationocene-natalie -aikens-amy-clukey-amy-k-king-and-isadora-wagner/.

Alexander, Michelle. 2010. *The New Jim Crow: Mass Incarceration in the Age of Colorblindness*. New York: The New Press.

Anand, Nikhil, Akhil Gupta, and Hannah Appel. 2018. "Introduction: Temporality, Politics, and the Promise of Infrastructure." In *The Promise of Infrastructure*, edited by Nikhil Anand, Akhil Gupta, and Hannah Appel, 1–38. Durham, NC: Duke Univ. Press.

Appel, Hannah. 2018. "Infrastructural Time." In *The Promise of Infrastructure*, edited by Nikhil Anand, Akhil Gupta, and Hannah Appel, 41–61. Durham, NC: Duke Univ. Press.

Bennett, Jane. 2010. *Vibrant Matter: A Political Ecology of Things*. Durham, NC: Duke Univ. Press.

Bone, Martyn. 2020. "Plantation Legacies, Environmental Racism, and the Anthropocene in Jesmyn Ward's Fiction." Paper presented at the Modern Language Association Annual Convention, Washington State Convention Center, Seattle, January 9.

Brogan, Kathleen. 1998. *Cultural Haunting: Ghosts and Ethnicity in Recent American Literature*. Charlottesville: Univ. of Virginia Press.

Buell, Lawrence. 1998. "Toxic Discourse." *Critical Inquiry* 24, no. 3: 639–65.

Bullard, Robert D. (1990) 2018. *Dumping in Dixie: Race, Class, and Environmental Quality*. New York: Routledge.

Carson, Rachel. (1962) 2002. *Silent Spring*. Boston: Mariner Books.

Davies, Thom. 2017. "Toxic Geographies: Chemical Plants, Plantations, and Plants That Will Not Grow." *Toxic News*, November 7. https://toxicnews .org/2017/11/07/toxic-plants-in-the-deep-south-chemical-plants-plantations -and-plants-that-will-not-grow/.

Del Principe, David, ed. 2014. "The EcoGothic in the Long Nineteenth Century." Special issue, *Gothic Studies* 16, no. 1.

Díaz, Junot. 2007. *The Brief Wondrous Life of Oscar Wao*. New York: Riverhead.

DuVernay, Ava, dir. 2016. *13th*. Los Angeles: Kandoo Films.

Estok, Simon C. 2009. "Theorizing in a Space of Ambivalent Openness: Ecocriticism and Ecophobia." *ISLE: Interdisciplinary Studies in Literature and Environment* 16, no. 2: 203–25.

Faulkner, William. (1950) 2011. *Requiem for a Nun*. New York: Vintage.

Fiedler, Leslie A. (1960) 1997. *Love and Death in the American Novel*. Chicago: Dalkey Archive Press.

Gan, Elaine, Nils Bubandt, Anna Lowenhaupt Tsing, and Heather Anne Swanson. 2017. "Introduction: Haunted Landscapes of the Anthropocene." Introduction to *Arts of Living on a Damaged Planet: Ghosts of the Anthropocene*, edited by Anna Lowenhaupt Tsing, Heather Anne Swanson, Elaine Gan, and Nils Bubandt, G1-G14. Minneapolis: Univ. of Minnesota Press.

Gilmore, Ruth Wilson. 2002. "Fatal Couplings of Power and Difference: Notes on Racism and Geography." *The Professional Geographer* 54, no. 1: 15–24.

Goddu, Teresa A. 1997. *Gothic America: Narrative, History, and Nation*. New York: Columbia Univ. Press.

Gordon, Avery F. (1997) 2008. *Ghostly Matters: Haunting and the Sociological Imagination*. Minneapolis: Univ. of Minnesota Press.

Haraway, Donna. 2015. "Anthropocene, Capitalocene, Plantationocene, Chthulucene: Making Kin." *Environmental Humanities* 6, no. 1: 159–65.

Haraway, Donna, Noboru Ishikawa, Scott F. Gilbert, Kenneth Olwig, Anna Lowenhaupt Tsing, and Nils Bubandt. 2016. "Anthropologists Are Talking–About the Anthropocene." *Ethnos* 81, no. 3: 535–64.

Harrison, Sheri-Marie. 2018. "New Black Gothic." *Los Angeles Review of Books*, June 23. https://lareviewofbooks.org/article/new-black-gothic/.

Hoberek, Andrew. 2007. "Introduction: After Postmodernism." In "After Postmodernism: Form and History in Contemporary American Fiction," edited by Andrew Hoberek. Special issue, *Twentieth-Century Literature* 53, no. 3: 233–47.

Keetley, Dawn, and Matthew Wynn Sivils, eds. 2017. *Ecogothic in Nineteenth-Century American Literature*. New York: Routledge.

Larkin, Brian. 2013. "The Politics and Poetics of Infrastructure." *Annual Review of Anthropology* 42: 327–43.

Lucas, Julian. 2016. "New Black Worlds to Know." Review of *The Underground Railroad* by Colson Whitehead. *New York Review of Books*, September 29, 56–57.

McKittrick, Katherine. 2011. "On Plantations, Prisons, and a Black Sense of Place." *Social and Cultural Geography* 12, no. 8: 947–63.

McKittrick, Katherine. 2013. "Plantation Futures." *Small Axe: A Caribbean Journal of Criticism* 17, no. 3: 1–15.

Misrach, Richard. (1998) 2012. *Ashland-Belle Helene Plantation, Acquired by Shell Chemical*. In *Petrochemical America*, by Richard Misrach and Kate Orff, 94–95. New York: Aperture.

Morrison, Toni. 1992. *Playing in the Dark: Whiteness and the Literary Imagination*. Cambridge, MA: Harvard Univ. Press.

Morrison, Toni. (1987) 2004. *Beloved*. New York: Vintage.

Oshinsky, David M. 1996. *Worse Than Slavery: Parchman Farm and the Ordeal of Jim Crow Justice*. New York: Free Press.

Pellow, David Naguib. 2018. *What Is Critical Environmental Justice?* Cambridge: Polity.

Rosen, Jeremy. 2018. "Literary Fiction and the Genres of Genre Fiction." *Post45*, August 7. https://post45.org/2018/08/literary-fiction-and-the-genres-of-genre-fiction/.

Sharpe, Christina. 2016. *In the Wake: On Blackness and Being*. Durham, NC: Duke Univ. Press.

Sivils, Matthew Wynn. 2013. "American Gothic and the Environment, 1800–Present." In *The Gothic World*, edited by Glennis Byron and Dale Townshend, 121–31. New York: Routledge.

Smith, Andrew, and William Hughes, eds. 2013. *Ecogothic*. Manchester, UK: Manchester Univ. Press.

Smith, Neil. 2006. "There's No Such Thing as a Natural Disaster." *Understanding Katrina Essay Forum, Social Science Research Council,* June 11. https:// items.ssrc.org/understanding-katrina/theres-no-such-thing-as-a-natural -disaster/.

Swanson, Heather Anne, Anna Lowenhaupt Tsing, Nils Bubandt, and Elaine Gan. 2017. "Introduction: Bodies Tumbled into Bodies." Introduction to *Arts of Living on a Damaged Planet: Monsters of the Anthropocene,* edited by Anna Lowenhaupt Tsing, Heather Anne Swanson, Elaine Gan, and Nils Bubandt, M1-M12. Minneapolis: Univ. of Minnesota Press.

Tompkins, Kyla Wazana. 2016. "On the Limits and Promise of New Materialist Philosophy." *Lateral* 5, no. 1. https://csalateral.org/issue/5–1/forum-alt -humanities-new-materialist-philosophy-tompkins/.

United Church of Christ Commission for Racial Justice. 1998. *From Plantations to Plants: Report of the Emergency National Commission on Environmental and Economic Justice in St. James Parish, Louisiana.* Cleveland, OH: United Church of Christ.

Ward, Jesmyn. 2017a. "For Jesmyn Ward, Writing Means Telling the 'Truth About the Place That I Live In.'" Interview with Terry Gross. *Fresh Air, National Public Radio,* November 28. https://www.npr.org/2017/11/28 /566933935/for-jesmyn-ward-writing-means-telling-the-truth-about-the-place -that-i-live-in.

Ward, Jesmyn. 2017b. *Sing, Unburied, Sing.* New York: Scribner.

Washington, Harriet A. 2006. *Medical Apartheid: The Dark History of Medical Experimentation on Black Americans from Colonial Times to the Present.* New York: Harlem Moon.

Whitehead, Colson. 2016. *The Underground Railroad.* New York: Doubleday.

Whyte, Kyle Powys. 2018. "Indigenous Science (Fiction) for the Anthropocene: Ancestral Dystopias and Fantasies of Climate Change Crises." *Environment and Planning E: Nature and Space* 1, nos. 1–2: 224–42.

Wright, Beverly, Pat Bryant, and Robert D. Bullard. 1997. "Coping with Poisons in Cancer Alley." In *Unequal Protection: Environmental Justice and Communities of Color,* edited by Robert D. Bullard, 110–29. San Francisco: Sierra Club Books.

Wright, Willie Jamaal. 2021. "As Above, So Below: Anti-Black Violence as Environmental Racism." *Antipode* 53, no. 3: 791–809.

Wynter, Sylvia. 1971. "Novel and History, Plot and Plantation." *Savacou* 5: 95–102.

Yusoff, Kathryn. 2018. *A Billion Black Anthropocenes or None.* Minneapolis: Univ. of Minnesota Press.

Kelly McKisson   The Subsident Gulf:
Refiguring Climate Change in Jesmyn
Ward's Bois Sauvage

Abstract This article focuses on figures of subsidence in Jesmyn Ward's novels of Bois Sauvage. Subsidence not only describes an actual process of sinking land in the US Gulf Coast bioregion but also refigures how those who study climate change can understand and address its material effects. A focus on subsidence makes visible the sometimes-invisible infrastructure of the ground, and analysis scaled to the figure of subsidence forces a reorientation of vision—away from rising sea levels and toward the destabilizing loss of land. From this perspective, Ward's fiction identifies histories of colonial engineering, extraction, and displacement as key ecological dangers. Unsettling national narratives of the Gulf Coast, Ward's subsident figurations connect issues of environmental emergency to structures of environmental racism, which unevenly enhance the precarity of certain communities by diminishing the ecological infrastructures of their lands. This article argues that literary fiction can produce new understandings of situated environmental challenges and can pose particular obligations for environmental justice.
Keywords Jesmyn Ward, US Gulf Coast, subsidence, figures, environmental justice

> The wind ripples the water and it is coming for us. . . .
> The snake has swallowed the whole yard and is opening its
> jaw under the house.
> —Jesmyn Ward, *Salvage the Bones* (2011)

Jesmyn Ward's *Salvage the Bones* (2011), a story of a family making their life in rural, coastal Mississippi, is now well known—and nationally applauded—as a novel of Hurricane Katrina. The storm builds as the story develops. For readers, who know where the novel is headed, the hurricane hovers over each turn of phrase. But when the storm finally arrives in the narrative and water rises on the family's land, the danger of the moment is figured as a discrete

American Literature, Volume 93, Number 3, September 2021
DOI 10.1215/00029831-9361279   © 2021 by Duke University Press

act of consumption: the "it" coming to feast on the family could be read as rising water, rippling in the storm winds, but it might better be read as the "snake," eating up the "whole yard" and "opening its jaw" to swallow the house from below (227).

This figure of a serpent swallowing up land from underneath those who live upon it is a refiguration of the hurricane event. Hurricane Katrina is typically remembered by the horrific storm surges, often in bird's-eye-view photographs. Indeed, rising waters drowned many in August 2005. Katrina was the deadliest hurricane to have hit the United States in over seventy-five years: when the storm made landfall on Louisiana, Mississippi, and Alabama, flooding along the coast ranged from ten feet above normal tide levels to twenty-eight feet above normal (NOAA n.d.). Water rushed onto land, filling in pockets of low and sunken earth and radically altering the topography of the landscape. Focusing on one community navigating that land, *Salvage the Bones* presents a ground-level perspective of this topography, a shift that refocuses readers' attention to the instability of the ground. *Salvage the Bones* emphasizes the swallowing up of land as the central tension, and danger, for those living in the Gulf Coast, in the path of ever-worsening storm systems. While the flood water is horrific, the loss of land beneath the characters' feet is even more terrifying.

Salvage the Bones produces a sustained figuration of sinking land, asserting a liveliness of environmental matter and, as Susan Leigh Star (1999: 377) puts it, "restor[ing] narrative to what appears to be dead"—the ground. Land is actively swallowed by the storm-snake in the moment of the hurricane and consistently unstable throughout the narrative, subject especially to the family's working it, their salvaging. The novel renders a lived, infrastructural experience of the changing environment. As Kregg Hetherington (2019) notes, the contemporary moment of the Anthropocene is marked by disturbances in distinctions, especially the distinction between environment and infrastructure. Infrastructure can refer to "the equipment, facilities, services, and supporting structures needed for a city's or region's functioning," as Patricia Yaeger (2007: 15) defines it. Or, ecological catastrophes may allow infrastructure to be realized as "the basic domains and cycles of the natural world: the global atmosphere, the water cycles and the waterways that it passes through, the soil and its fertility," as Jedediah Britton-Purdy (2018) suggests. Britton-Purdy argues that "these, too," water systems and land structures, "are infrastructure: the conditions of all human action and interaction. Human activity increasingly shapes them and will do so even more intensively."

Ward's contemporary novels are crucial to discussions of climate change because they perform a kind of inversion, directing readers to notice the instability of the ground and reminding readers that "infrastructure only recedes into the background for those who are not busy building or repairing" (Hetherington 2019: 6).[1] Yaeger (2007: 17) suggests that "when infrastructure disappears . . . we should remember to read this absence as a taking for granted of infrastructural privilege." Ward's figurations of sinking land make apparent for readers and scholars what has been naturalized as given: reliable ground upon which to live is required infrastructure for making life. This critical inversion flips the perspective of analysis from the threat of flooding waters to the precarity of unstable lands. While climate change discourse reports on the environmental dangers of rising global temperatures and sea levels, Ward's narratives embedded in grounded ecologies of the Mississippi delta and Gulf Coast encourage scholarship to attend to the complexities of specific geophysical infrastructures— not all coasts are equally, or in the same way, at risk.[2]

All three of Ward's novels, *Where the Line Bleeds* (2008), *Salvage the Bones* (2011), and *Sing, Unburied, Sing* (2017), are set in her fictional Mississippi bayou town, Bois Sauvage, and provide a ground-level perspective of the US Gulf Coast, registering global changes as relative to the dynamic surfaces of local delta lands. This embedded perspective is able to highlight how the infrastructure of the Mississippi delta system is not only subject to "global mean sea-level rise" but also to fluctuating "rates of fluvial sediment deposition" and "accelerated subsidence" (Ericson et al. 2006: 65). This distinction is significant: the Gulf Coast is not just subject to rising waters but also, and particularly, to subsidence. Land here has been disappearing—not just eroding but also sinking—at a rate of about an acre an hour for an estimated total of about two thousand square miles lost in the last century.[3] The reorientation to sinking and disappearing lands connects the issue of global climate change to the intimate realities of Gulf Coast land management, processes of river engineering and resource extraction that continue to accelerate the natural rates of subsidence.[4]

Ward's focus on sinking land as failing infrastructure develops an aesthetics of subsidence that not only describes an actual process experienced by those who live at ground level but also refigures the way those who study environmental change can understand and address its embodied realities. Donna J. Haraway (1997: 11), who takes seriously figures as method, notes that "figurations can be condensed maps of contestable worlds."[5] Figuring is powerful for Haraway because it can perform a remapping or reimagining of the relations of our world. Judith Madera (2015: 23) shows how African American literature has long "perform[ed] the work of refiguration" to "open[] vital geographies" for Black life. Christina Sharpe's (2016: 18) critical work uses "wake" as a metaphoric figure to remap how Black lives are conditioned by the afterlives of slavery and "to imagine new ways to live . . . to survive (and more)." With the gathering of the many meanings of "wake," Sharpe aims to "sound a new language" and reimagine care for Black life (19). My analysis of subsidence in Ward's work aims to show how these novels refigure issues of environmental catastrophe as necessarily also issues of environmental justice. Subsidence, a term describing geophysical processes of the US Gulf Coast bioregion, here also aims to offer new language for how global climate change coevolves with ongoing histories of colonial dispossession and expropriative capitalism to pull the ground out from underneath Gulf Coast communities.[6] Subsidence remaps the lived experiences of climate change in the Gulf Coast as not simply naturalized vulnerability to rising sea levels but rather as a complex, historical and biopolitical precarity in relation to sinking ecological infrastructures.[7]

This essay gathers Ward's subsidence figures to show how the experience of climate change is refigured in her novels of Bois Sauvage. These novels can be read as documenting a climate change history: they represent extreme weather events, notably Hurricane Katrina, and depict ecological catastrophe, the consequences of the Deepwater Horizon disaster. But this essay aims to show how literary fiction can do even more to respond to lived climate change realities.[8] Ward's novels refigure isolated events of environmental destruction as connected legacies of environmental violence. The figurative aesthetics of subsidence that I examine illustrate how the precarious infrastructure of the US Gulf Coast and Mississippi delta is marked by a metaphorical and literal deplacing—both the historical displacement of many peoples, especially indigenous and incarcerated peoples, repeatedly forced out of their homes, communities, and lands and the ongoing displacement (in place) of those whose land is disappearing. In the delta and Gulf Coast, solid ground is shifting; the land is submerged in the swamps and bayous, inundated at the levees and eroding coastal shores, and sinking from extraction and degradation.

Literary figures of subsidence, as I track them through Ward's novels, can both produce new understandings of these situated environmental challenges and pose particular obligations for environmental justice.[9] This essay begins with *Where the Line Bleeds* (2008) to demonstrate how the post-Katrina absence of aid to Bois Sauvage is figured as unsettling characters who sink on ground that can no longer support their subsistence. The economic and environmental challenges here are particularly racialized, and the degrading infrastructure cannot be separated from ongoing histories of environmental racism. *Sing, Unburied, Sing* (2017) demonstrates the geographical reach of Ward's figurative powers, upending the traditional North-South imaginary as characters move deeper north and further into danger. Even though this narrative takes place on the move, we see the characters immobilized and swallowed by haunting histories. In this novel, water and the coastline function like a reprieve from the subsident pull of the land. In *Salvage the Bones* (2011), the threat of storm surge is overshadowed by the continuous loss of the land on which a family makes their life. Here, readers can identify as primary violence the swallowing up of ground that underwrites the hurricane's consequences. I end with *Salvage the Bones* because it most explicitly presents a resistant strategy to the consumptive violences unsettling the infrastructure of Bois Sauvage. This novel takes joy in salvaging as a form of life-making required for existence in this subsident place.

Before turning to the novels, it is important to emphasize how these aesthetic figurations of ecological infrastructure productively respond to the real, lived experiences of ecological violence. Subsidence is a key analytic for these novels because subsidence is central to the infrastructural precarity of subsisting in the US Gulf Coast region. Perhaps counterintuitively, land here is made and remade by the waters travelling down the Mississippi. The Mississippi River, the longest in North America, carries silt and sediment from about one-eighth of the continent, collecting from at least thirty-one US states and two Canadian provinces.[10] At the mouth of the river delta, where the Mississippi meets the Gulf of Mexico, sediment is deposited and levees are formed, extending and augmenting the grounds of the wetlands. In waves of settlement and colonization, human activity has amplified this process by building up the banks of the river through what is called artificial leveeing and by cutting channels through the marsh for more direct access to the Gulf Coast waters. The artificial levees that provide sanctuary to human settlements near the banks of the river block more than just flood waters; they also interrupt the process of sedimentation that continuously rebuilds the outlying lands of the bayous, effectively disappearing those lands. The channeling—cutting direct pathways through the complex, webbed bayous in order to run ships and pipeline to the mainland—disrupts the chemistry of the wetland environment, which leads to ecological die-off and erosion. Beneath these reworkings of topography are energy-industry wells that, through forms of extraction, cause underground shifting with the effect of surface-level sinking.[11] Leveeing, channeling, and extraction, often acts of national and international origin, perform particular and intended restructurings of the ecological infrastructure that effect an often-invisibilized deplacing for local inhabitants. It can be difficult to trace how the long and slow histories of colonial and capital restructurings are directly responsible for discrete, often-sudden acts of violence upon peoples living here.

Imaginative and aesthetic figures can be vital tools for visibilizing the abstract connections and transformative violences of ecological damage. For example, in an uncanny echo of Ward's snake figure, Antonia Juhasz's (2013) contribution to Rebecca Solnit and Rebecca Snedeker's atlas project of New Orleans, *Unfathomable City*, figures the history and infrastructure of the Gulf Coast energy industry as a tentacled and consuming monster sunken into the basin. Immediately following an image that maps a flat network of oil wells, platforms, pipelines, refineries, and shipping lanes, Juhasz's (2013: 53) description brings the infrastructure to life: "The Great Gulf Oil Monster is literally devouring the coast as spills intensify the damage." Juhasz imagines the transnational, billion-dollar oil and gas industry as a creature with "thousands of steel-pipe tentacles reaching out," distributing poisons "through the ocean like a writhing sea monster" (50 and 51). This figuration makes strange what may seem, by today's standards, simple and typical energy infrastructure; to call such structures monstrous is to make evident the danger involved in what the industry considers "usual" operating procedure (51). Juhasz's description of "the monster's sheer weight, which brought the noxious matter to the ocean floor where it rests today," figuratively demonstrates how the effects of the industry linger. The "sheer weight" of the monster settling and "writhing" imagines how capital, as Stephanie LeMenager (2014: 13) describes it, "bulks out and inhabits place, changing the quality of air, water, noise, views, and light." The creature persists, not only by settling into the region that economically depends on its business—thus, haunting the peoples that live there with its pervasive presence—but also by "devour[ing]," a monstrous consumption that feeds upon the land, eating it up from under those who must live upon it.

The "Gulf Oil Monster" demonstrates the need to understand ecological infrastructures of the Gulf Coast in more than two dimensions: critical analysis must be able to think vertically, with depth, as well as through time.[12] Solnit and Snedeker (2013: 11) call New Orleans "a bowl that nothing ever leaves" to represent the way that the place accumulates systems and meanings that then linger by sinking. Sinking in a bowl is a subsident figuration depicting the region as not only comprised of flat areas, nationally bounded by longitude and latitude, but also fleshed out in a third dimension. Stephen Tatum (2007: 4) proposes that we use "overlaying or doubling" to understand how global processes (international flows of capital and commodities, new information technologies, migration of labor and people) restructure places, "redrawing and determining the contour lines of the local, the regional." Tatum's paradigm of spectrality describes the way subsidence forces a three-dimensional perspective; he imagines place as "bubbling over" with material and abstract presences (11). This perspective makes sense in a place that is often literally bubbling over, where land structures change constantly with flows of river and gulf waters and flows of commodities and labor. Figures such as "bubbling over" also collapse time by articulating in one moment a process that occurs on a longer time scale.

The figurative power of literature can uniquely remap across space that is not so much topographical as topological, bending and twisting independent of size and shape as narratives move through their own time.[13] Ward's novels of the US Gulf Coast reject the dominance of national geography, framing the narratives instead by "the relations that come together to make place" (Madera 2015: 13). These fictions follow the intimacies of this imagined place, Bois Sauvage, situated at the level and scale of the human (and often animal) bodies that make up its community.[14] The narratives move "both centripetally *and* centrifugally," to remap and refigure in greater complexity the perspective of Black families living on the land (Crownshaw 2016: 230). Subsidence figures generate sinking anxieties and upend orientations to convey the claustrophobia of being swallowed. The situated perspective of sinking earth, of subsidence, remains stuck in the trouble of the embedded human perspective and encourages investigation at a more intimate scale.[15] From this perspective, and in this representational mode, readers can understand the history of colonial engineering and extraction as key ecological dangers acting at the level of the body. By refocusing to acts of dispossession and displacement and revaluing the function of water, Ward's novels clarify the imperative for justice. Instead of calling for a shoring up of national borders with levees and walls, subsidence figures ask us to consider structures of environmental racism that unevenly enhance the precarity of certain communities by diminishing the ecological infrastructures of their lands.

Ward's 2008 novel *Where the Line Bleeds* is set post-Katrina, in the

wake of flooded destruction: echoes of Katrina slip into the text generally, through the town's lack of industry, increase in drug traffic, and
need for repairs, and more directly, written on wood boards collected
for the next storm, plywood "marked hurricane" (213). But rather
than focus on the "natural" catastrophe, the novel redirects readers'
focus to the particular, pointed dangers trawling for our main characters, like hooks dangling in the water to trick unsuspecting fish. The
novel's subsidence figurations illustrate how unemployment and poverty are responsible for trapping characters without means of subsistence. Joshua and his brother Christophe, twins and main characters
of the novel, are recently graduated from high school, out of work,
and fast running out of money. The pressure to seek income keeps
them stuck and sinking in a spiral of waking up, leaving home to fill
out employment applications, then returning empty-handed, sweat-
ing in the humid heat of the day and in the knowledge they will have
to start over again the next morning. Joshua equates this feeling to
shrimp caught in a crawler, "struggl[ing] against the thick fingers
of the current created by the encroaching net in hope of escaping, of
moving forward" (71). The water is the shrimp's ecological home, and
it is the intruding crawler that creates the devastating current, that
suffocates with its net.

In this first novel Ward most clearly maps Bois Sauvage as a place
characterized by a subsident pull. *Where the Line Bleeds* maps the
space through a lived geography and history, bringing the town to life
as a character:

> Bois Sauvage dug in small on the back of the bay, isolated. Natural
> boundaries surrounded it on three sides. To the south, east, and
> west, a bayou bordered it, the same bayou that the Wolf River emp-
> tied into before it pooled into the Bay of Angels and then out to the
> Gulf of Mexico . . . To the north, the interstate capped the small
> town like a ruler, beyond which a thick bristle of pine forest stretched
> off and away into the horizon. It was beautiful.
>
> Joshua could understand why Ma-mee's and Papa's families had
> migrated here from New Orleans, had struggled to domesticate the
> low-lying, sandy earth that reeked of rotten eggs in a dry summer
> and washed away easily in a wet one. Land had been cheaper along
> the Mississippi gulf, and black Creoles had spread along the coast-
> line. They'd bargained in broken English and French to buy tens of
> acres of land. Still, they and their poor white neighbors were depen-
> dent on the rich for their livelihood. (6)

Readers see the Bois through Joshua's eyes. He is mesmerized, in awe of the land—"It was beautiful." Rowan Jacobsen (2011: 5), writing about the US Gulf Coast, provides a theory of why humans are drawn to the coasts: "Not only because we like gazing out at the rolling waves and the sea of infinite mystery and possibilities beyond, but also because the coasts of the world are uniquely abundant in everything that makes our lives grand." However poetic, Jacobsen's claim is ecologically sound—interstitial coastal zones, such as the wetlands, estuaries, marshes, and bays at the northern rim of the Gulf of Mexico, are nutrient-rich, protected environments that provide habitat for a great variety of life. Jacobsen himself keeps coming back to the Gulf Coast, drawn to the same beauty Joshua might see, but Ward's novel provides simultaneous, and alternative, realities of life in the Gulf Coast: "Land had been cheaper" here, Joshua's family memory speaks back; Black Creoles speaking "broken English and French" alongside "poor white neighbors" bargain and serve out of necessity; people struggle to make a life on the land that no one wanted, that is unstable, and that "washed away easily." Where Jacobsen imagines possibilities, the lived experience voiced through the novel presents a subsident precarity: water borders the town on three sides and produces a sinking motion, emptying and pooling south, into the bay, then into the gulf. At the fourth border, to the north, is an interstate, a transit line that might represent a way out but here is figured as another threat, keeping a people pressed in, "capp[ing] the small town."

An apt description for the plot's conflict is not rising action but instead suffocating anxiety. When Joshua receives a callback from the docks—the only well-paying labor in town—the pressure increases for Christophe, materializing as a dark fear that surrounds him, "the black tree limbs suddenly seemed like fingers . . . something was closing in on him" (Ward 2008: 60). The text hints that it is not by chance that Joshua is called back for a job at the dock while Christophe is not. Employers are more attracted to Joshua in the way that Christophe understands their friend Laila is more attracted to Joshua— for "his light eyes, easier-to-braid hair" (94). Joshua has lighter— coded whiter—features, and the twins' subtly different racialized appearances manifest when it becomes clear that Christophe has to work harder to stay afloat than Joshua does. Christophe, in the wake of a darkness beginning to pull against him, notes that he should make himself look neater for his interviews as "prospective employers would think him lazy and unreliable if he wore his hair wild and curly" (94). Here in Bois Sauvage, where good jobs are so hard to get, the racially discriminating processes of capital drive a wedge between the boys, setting them both on edge.

Unsettling anxieties accumulate and call attention to the failure of the town's ecological, and thus economic, infrastructure to support its community. Floods have destroyed properties and damaged homes of families who are without the income to build anew and whose property values have long been pulled down by redlining practices. Absent from the text is any centralized aid or state assistance—whatever economic boom resulted from reconstruction is not present for our characters, who do not see the tax dollars shuffled toward private contractors. Private industry appears via drilling and toxic spillage that is just off the edges of the novel but that pervades the gulf waters and destroys marine ecosystems, especially damaging the fisheries. Readers see how the industry of this place fails to support its inhabitants, leading them, like Christophe, to choose the drug trade, or drug use, to try to make a life.

The figuration of ecological instability as failing infrastructure makes visible a legacy of slow violence that has been waiting to hit the twins—the novel depicts their coming of age as an inheritance of that legacy, which has been pulling down Bois Sauvage for some time. Christophe ends up selling weed, an unsteady job that he knows feeds the "inevitable" cycle of "jail and hustling" (56). Since the coast is "too small" for dealers to settle down, they end up displaced, "running, hiding, haunted" (56). Ma-mee, the boys' grandmother, describes the drug addicts of the town as "unsettled," "perpetually waiting for something astounding to happen: a tornado, a flood, an earthquake" (89). This anxiety figuratively links the extended temporality of addiction to the recognizable immediacies of ecological catastrophes. The "unsettled" remain "perpetually waiting" because they are merely reminders of a change that has already taken place. The novel shows that displacement is a cross-generational event already in process, figured through brief glimpses of the father, importantly named Sandman, who sinks in and out of rehab, and then through Christophe, heir to this subsident earth.

The subsident legacy of Bois Sauvage is a spiral of diminishing returns, increasingly unable to support younger generations who might have sustained the cultural traditions. Ma-mee functions as the novel's historical memory, and through her readers see how changes to this community are figured as disappeared losses. Ma-mee registers the movement of the drug trade into town as a presence accentuating the absences of her cultural infrastructure. The relations maintaining Ma-mee's creole heritage and community have been expropriated and deplaced by larger forces of capital. She is alone in remembering what used to be here: the place of her childhood, where she gathered "Spanish moss with her mother," is lost to her (64). She catches herself "dream[ing] in a language that no one around her spoke any longer," her knowledge of creole French reminds her, like her "wide, lonely bed," that she, too, is growing old and fading (64). Now isolated, unable to maintain her culture as those around her have died or moved elsewhere to survive, Ma-mee feels "fits of lethargy and utter exhaustion" as "a shadow passing over her," her blindness accentuating the image of being buried alive (65).

If the unsettling racialized treatment and the seductive pull of drug use are subsident forces, in *Where the Line Bleeds* water becomes a means of orientation for people of the Bois, both sanctuary and opportunity. "Joshua love[s] the coastline" (29), and he finds brief moments of release looking out over the water or at the gulls sailing over the docks (90). Driving, especially along the coastline, is a kind of freedom; when Christophe observes the marsh through the car window, he identifies the familiar scene as offering a clear, if imaginary, way out from the claustrophobia: "the water . . . a dark blue, the reflection of the moon shimmering like a white stone path on its surface" (49). Whereas land is more mercurial, not to be trusted, water is a constant, comforting feature: Joshua is eager to "feel weightless and buoyed" (71). Water is an anchor for the characters' lives and the plot, which begins and ends at the water: the boys jump off the bridge into the river on the first page and are fishing with cousin Dunny at the river on the last. In the final passage of the novel, the boys are figured as fish, thrown back after being caught on the line. Readers may choose to be comforted by the idea that, even if they seem to be unable to rise up in some narrative progression toward success, the boys can float without sinking, buoyed by the gulf waters.

Rather than force characters simply to bear the weight of oppressive histories of displacement, enslavement, forced migration, and incarceration, Ward's novels explore the way characters resist that sinking pull by reimagining their relations to land. *Sing, Unburied, Sing* (2017) takes place in the same fictional Bois Sauvage mapped in *Where the Line Bleeds* and ends with a similar image of compromised freedom, two of the main characters experiencing escape as they drive through the town. Here, the characters refigure their world as "an aquarium," an escape of their own making (Ward 2017: 274). More elaborately than in *Where the Line Bleeds*, *Sing, Unburied, Sing* emphasizes the characters' abilities to remap and reimagine their geological and geographical space. The remapping is enabled by emphasizing the subsident depths as a layered space of haunting histories. *Sing, Unburied, Sing* tells the story of a Black child living in the Mississippi delta, a story interwoven with his white father's memories of the Deepwater Horizon disaster, his Black mother's navigation of their journey north to the heart of Mississippi, his grandfather's retelling of . Jim Crow–era imprisonment, and a ghost child's revelation of his terrifying lynching. Rotating narrators and overlaid timelines construct a cyclical, thick narrative moving deeper into this place.

Though plotted almost wholly on the road, *Sing, Unburied, Sing*'s narrative feels forever stuck and sinking. The novel is set mostly in a car where Jojo, the thirteen-year-old protagonist, rides in the backseat with his sister, Kayla, while their mother, Leonie, drives them north, away from their home in Bois Sauvage. They travel toward Parchman Prison, the Mississippi State Penitentiary where the children's father, Leonie's boyfriend Michael, has been imprisoned for a drug-related conviction. The narrative space of the car ride consistently sinks into memories and dreams, rooting the characters even further in the South as they move north and routing them by the quicksand-like histories of slavery and dispossession. This narrative stretching, up and down as the plot moves north and south, provides space for the novel to show overlapping legacies of violence lingering on the land and conditioning the characters' lived experiences.

As in *Where the Line Bleeds*, the subsident gulf land is not a home that promises security or stability—Leonie's drowning dream figures the failing infrastructure as a raft that bursts and leaves her whole family at risk: "We are all sinking . . . they won't stay up: they want to sink like stones" (195). The subsident land in which the characters are embedded figures not only their anxieties but also their embodiment. Mam's body is likened to marsh land sinking under the patterned rise and fall of the tide water by the cancer that came and left and came back again (26). The image of sediment in water describes how the family "gift" (an ability to hear things talk—animals, plants, sicknesses—and see things that are not real) runs generationally through the family blood, "like silt in river water" (40). Murky quicksand is the grief Leonie "[swims] up and surface[s] out" of when she's with Michael (153). Michael's skin is like water over land, tepid to touch as Leonie lays her face along his arm (154); when Michael is using, it strains against cheekbones that stand out "like rocks under water" (93). When Mam dies, she floats away as "time floods the room like a storm surge" (269).

Water figures reparatively in this novel, too, as a sound, a complement to and then a refiguration of the song that relieves the ghosts: "*Home*, they say," repeating the "the sound of the water in Leonie's womb, the sound of all water" (285). Throughout the novel, rather than an encroaching threat, water is a helpful compass. Water is how the characters orient themselves as they move north across the state, and water marks the separation of their safe coastal Bois from Parchman Prison. Pop's retellings of Parchman remind readers how far north it is from the Gulf Coast, from the water: "It's different up there. The heat. Ain't no water to catch the wind and cool you off" (22). It is so important that he repeats it: "Down here, it's different; we got that wind coming off the water all the time, and that eases. But up there, they ain't got that" (119). The prison's very name indicates a harsh, material reality, and Jojo thinks the connection for us: "I wonder who that parched man was, that man dying for water . . . Wonder what that man said before he died of a cracked throat" (63). As the car creeps north, page after page, the drive gets hotter and drier and the passengers keep getting thirsty. The tension is heightened as Kayla becomes progressively ill, running a fever and vomiting. As the car heads back toward home, the danger of being Black and on the move in Mississippi is made violently explicit: police officers stop the car and literally pull the characters down to the ground. It is a relief when the car finally comes home to the Gulf Coast, as if crawling out of some sunken place into the fresh, cool air.

When *Sing, Unburied, Sing* refigures water as offering a source of recovery from the unstable ground, it also remaps the relations of North to South. The plot marks the North as dangerous, imprisoning land, while the South is likened to coming up for air. This is an upending reorientation of the historical narrative division of "Northern freedom and Southern bondage" (Madera 2015: 11). The novel flips this story, retelling and thus remaking the map to freedom for the characters. Madera (2015: 4) theorizes the way African American literature creates worlds that redraw official maps as a process of "deterritorialization" that can "find openings for different forms of actualization." For Black authors, and Black characters, Madera explains, this can mean "writ[ing] over white principles of containment" and "dismantle[ing] dominant organizational codes of place" (3). We might be tempted to read this positive movement farther south into rural spaces of the Gulf Coast, further removed from the arms of the central government, as resistance to the biopolitical structures of the United States. Nonetheless, one important figuration in the novel reveals that even here submerged systems of capital organize life.

Near the end, *Sing, Unburied, Sing* discloses another sunken history and reminds readers that characters can be imprisoned by more than buildings: as insatiable forces of capital extend into the gulf water seeking sources of accumulation, private interests impoverish and poison local resources. Ward engages an extended figure of subsident spiraling to clarify the consequences for individuals trying to make their lives on such failing ecological infrastructure. Upon return from Parchman, Michael continues to be impacted by nightmarish memories of the Deepwater Horizon disaster. Michael was working as a welder on the rig when the explosion happened, and he relates his experience to Jojo:

> [Michael] told me about working out on the rig. How he liked working through the night so when the sun was rising, the ocean and the sky were one thing, and it felt like he was in a perfect egg. How the sharks were birds, like hawks, hunting the water. How they were drawn to the reef that grew up around the rig, how they struck under the pillars, white in the darkness, like a knife under dark skin. How blood followed them, too. How the dolphins would come after the sharks left and how they would leap from the water if they knew anyone was watching, chattering. How he cried one day after the spill, when he heard about how all of them was dying off. . . .
>
> *I actually cried*, Michael told the water. He seemed ashamed to say that, but he went on anyway. How the dolphins were dying off, how whole pods of them washed up on the beaches in Florida, in Louisiana, in Alabama, in Mississippi: oil-burnt, sick with lesions, hollowed out from the insides. And then Michael said something I'll never forget: *Some scientists for BP said this didn't have nothing to do with the oil, that sometimes this is what happens to animals: they die for unexpected reasons.* And then Michael looked at me and said: *And when the scientist said that, I thought about humans. Because humans is animals.* And the way he looked at me that night told me he wasn't just thinking about any humans; he was thinking about me. I wonder if Michael thought that yesterday, when he saw that gun, saw that cop push me down so I bowed to the dirt. (Ward 2017: 225–26)

The "perfect egg" of Michael's contentment reveals a complicated position toward the oil rigs: something was pleasing about this job, which provided income. The proximity to water is figured as a sort of womb making life possible. The rig itself, in Michael's recollection, becomes substitute ecological infrastructure—the flourishing reef, the hunting sharks, the gregarious dolphins—and it causes him grief to know this is dying, even in its violence. As the description moves deeper, below the rig, the violence is revealed to be a specific predation, "white in the darkness, like a knife under dark skin." The threat here is against racialized Black bodies, and it compromises the perfection of the womb ecology and foreshadows dangers to come. When Michael thinks through the implication that humans, too, will be harmed by this disaster, he thinks of certain humans; the novel aligns environmental damage and police brutality as similar kinds of racialized structures that direct the uneven operation of power. Here, the novel's strange figuration of the rig, the dying animals, and police violence against Black bodies forces a reader to hold these seemingly disparate elements together at one time and forces a line to be drawn down, connecting one to the next in a kind of spiral, a new map of what it is like to try to live here. Michael's remapping makes clear, in his skeptical retelling, that the destruction—pods of dolphins washing ashore, bodies toxic and rotting out—is not just *"what happens to animals"* any more than the police brutality inflicted against Jojo is not just what happens to people. These actions are made unnatural in Michael's telling and linked together as violences inflicted by similar causes.

The ecological infrastructure of Bois Sauvage provides Michael with few options for economic and social stability. Failing to overcome his depression and grief and rooted in a town where he cannot find work, Michael remakes a way by drug sale and use, the only route through which he sees possibility for himself and his family. The narrative relays this route in a moment of palimpsest; Leonie reflects on Michael's past at the same time as she looks forward toward that same reality (91–93). The two experiences are compressed into one point, for Leonie is running meth up to the prison on their journey north to retrieve Michael in order to make possible that very journey. The only way forward at the same time reinscribes both characters back into the sinking path that keeps the whole family stuck in place.

Michael and Leonie's attempt to remake their possibilities is part of Ward's larger attention to salvaging as a dominant strategy for everyday survival. Salvaging involves a continuous remaking of subsistence conditions in the face of failing infrastructures such as subsident land. In *Salvage the Bones* (2011), Hurricane Katrina frames a narrative of making life in the Gulf Coast region, where the land is sinking under water. The novel reveals that the ecological infrastructure is made unstable and the bodies on it made fungible by the entwined histories of enslavement and extractive capitalism. Subsidence connects histories of displacement to conditions of environmental injustices figured by forces that unmake worlds: childhood pregnancy, poverty, and powerful storms. Here, the tragedy of the storm and the havoc that will follow has been preordained by the economic inequalities the community persists through every day. Katrina is depicted as the continuation of a longer legacy of oppression figured by land destruction. The salvaging acts of the characters are recoded as acts of dissension, resistances to the subsident forces that would swallow them down into the sinking delta earth.

Salvaging is a response to subsidence that orients the characters and the reader to the racialized realities of making life on failing infrastructure in the Bois. Either salvaging is a practice of bricolage, the characters making due with what they have and can find to counter instability, continuously borrowing from homes for scrap or remaking those homes with pieces of surrounding structures; or, salvaging is stealing, a crime with fearful consequences. The town is bisected along racial and class boundaries that govern this binary logic: as the characters move from their familiar "black Bois" into the "white Bois," the narrative recalls histories of racial segregation and neglect (Ward 2011: 70). Movements into the white Bois are always tense, constrained, and in darkness, while relief is found back in the Black Bois. Though white homes are nicer, better maintained and stocked with materials that the characters can use to rebuild their homes and care for their families, salvaging in the white Bois becomes theft. The characters prefer the Black Bois, the "black heart of Bois Sauvage," to the "pale arteries" of the white parts of town (97). The fact that the Black Bois is more fluid, more unstable, than the white Bois aids the characters' salvaging practices.

Salvaging relies on this complicated relation to unstable infrastructure: the most valuable resource this community of primarily Black families has is the land itself. Fungible, the land has been, and continues to be, bartered off for sustenance. The "Pit" is the name of the land that the protagonist, Esch, lives on with her family, and "pit" in the novel becomes a primary figure of subsidence that exposes the consumptive forces feeding off of the family even before the hurricane lands. Formed originally because the grandfather, Papa Joseph, had only the land under him to sell—so, he "let the white men he worked with dig for clay"—the Pit is an explicit marker of histories of disenfranchisement and poverty that literally pulled the land out from under communities in the region (14). Over time, the absent land of the Pit formed a bowl and a boundary to the property, a sunken pit that the children explore and that the family uses as a dump. When it is dry, the pit is merely a dump, yet, when it rains long and hard enough, the pit fills up with water and becomes a swimming hole for the children to play in (15).

By focusing on the characters' remaking through play and material refurbishment, the novel depicts how the instability of the land can be countered by the peoples' resistant and imaginative postures of dissension. This is an important function of literature, as Madera (2015: 4–5) explains, giving shape "to spaces that go underrepresented in traditional cartography"—such as the Pit of Esch's family land— "mak[ing] contours for spaces of dissention . . . yield[ing] new forms." And, as with Ward's other novels, water plays an important role in the characters' ability to imagine. The coming hurricane is welcomed by the children, who anticipate rain for good swimming water. The storm also spurs movement in the father, who begins fortifying the house and readying the tractor. Though the hurricane warns of effects on the house, it also promises work: "They going to be money to be made after this storm come through," Daddy predicts, "by a man with a dump truck" (Ward 11: 90). Like the ebb and flow of the tidal rhythm, Esch and her family continuously remake the land and the structures around them into a place, necessarily also temporary and compromised, that they call home.

Dissention through salvaging is one strategy for facing failing infrastructure, and it is represented as an assertion of life. As the novel begins with labor, it connects one futurity represented by the birth of Skeetah's dog's puppies to another, the process of Esch's father "fixing up" the house to ride out the storm; both are means of yielding new forms (4). Just like the pain and struggle of the dog, the father's labor requires more tearing apart than might be assumed. The family makes use of materials around the property by breaking things down in order to build up the shelter of the home. In this way, too, salvaging echoes the path of the Mississippi, the way the river pulls soil loose from one place and transfers it hundreds of miles downriver to shore up the coastal lands. However, the novel reveals how this way of securing safety can have contrary results. If the father's plan is likened to the production of artificial levees preparing for the storm surge, his strategy has already failed. If breached, the levees can retain water on land that has been displaced, holding water inside the walls and flooding the sunken ground. It turns out that rising waters pose a lesser threat than failing infrastructure.

Given this far-reaching subsidence, when the family follows the father's guidelines to prepare for Katrina, they cannot overcome the disadvantage of the Pit. Instead of a fortified refuge, the Pit becomes evidence of legacies of mistreatment, and the hurricane filling up the pitted land reanimates the consumption of Black bodies by "the hungry maw of the storm" (230). This is what happens when Katrina hits: the world that the novel has built, the unstable Pit, allows water to seep in, and "like a creeping animal," the figure seems to have aim and intention—"it is coming for us" (227). But it is not the water that is coming for the family. It is a different, figurative agent, a snake, a creature with an ever-widening jaw unhinged to eat things greater than itself. As the surge fills up the house, drowning is rendered as a form of attack. This land and these bodies become fodder for consumptive structures of colonialism and capital. The weighty fact that the family members are consumables grows as the storm brews, each accretion of narrative tension weighing down the plot: The puppies die, one by one, increasing the affective pull of depression. Esch feels the increasing weight of the growing child inside her, her narration increasingly exhausted by morning sickness and aching joints. Skeetah, his dog, and the father receive serious wounds. The father's dismemberment stops not only his activity but also the family's hope for mobility and profit by machine labor. The novel remaps the path of the storm through these plot points, which pull the story deeper and deeper into figurative and literal darkness, preempting the climax of the storm. Rather than from the bayou, "where we thought the water would come from, the reason we thought we were safe," the danger comes up from under the characters as they lose figurative and literal ground (248).

Even in the wake of emergency, the work of making life is required. Ward's fiction does not only portray victimized, racialized bodies sinking under the weight of ecological infrastructural failure. Rather, as Rick Crownshaw (2016: 225) notes, "Ward's emphasis on the ideas of survival and renewal—a 'savage' resilience of humanity in its most precarious state—offers a corrective to the proclivities of some critical theory deployed in the framing of Hurricane Katrina's victims and the longer history of suffering they represented." Ward's novels figure life in the rural, coastal Bois as unsustainable, but the characters' salvaging strategies make subsistence seem possible. In the aftermath of the hurricane, Esch must remake her own life even as she prepares to produce new life. Looking out over the Bois to survey Katrina's rearrangement of the town, Esch ponders possibility, wondering "what the storm has stirred up from the bottom of the bay, and what it has dragged in and left in the warm, mud-dark water," presumably for her to find and use in her remaking (249). Ward's characters continuously remake paths across subsident earth. Besieged not only by forces of climate change but also by structures of national and international capital with attendant raced and gendered violences, these characters make lives on unstable ground by remaking, remapping, and refiguring place.

Ward's novels of Bois Sauvage foreground ecological infrastructure by situating embodied experiences within long histories and in relation to larger-scale forces. Aesthetic figurations of subsidence are central to this work, enabling readers not only to attend to the changing environment but also to make connections across space and time to legacies of colonial violences and capitalist expropriations. Ward's figurative language engages the movement of subsidence by narrativizing unstable and sinking experiences of ecological infrastructure in the rural Gulf Coast. This ground-level perspective offers insight into a deep history of environmental injustice, forcing a vision distinct from that of an abstracted, global imagination—neither above the blue marble nor surveying a flat map. In fact, the novels are able to refigure a dominant focus of climate change criticism: sea levels are rising, but the emphasis on this aspect of change often limits the ways we might imagine possibilities for salvaging life on land.[16] As I've gathered them here, subsident figures in Ward's novels importantly reorient our ecological attention to the sinking lands disappearing in the Mississippi delta. This perspective opens up ecocriticism to questions of dispossession, environmental racism, and capital accumulation. Ward's figurations pose environmental challenges as issues of environmental justice, calling readers and scholars alike to consider justice as our primary responsibility in discussions of global climate change. Literature, here, has the power to reframe instances of ecological and infrastructural emergency as opportunities to salvage more just futures.

Kelly McKisson is a PhD candidate in English at Rice University. Her dissertation focuses on weird ecological figures in contemporary American literature.

Notes

My deepest thanks to Krista Comer, for her invaluable encouragement, and to Rosemary Hennessy, for her generous feedback on drafts of this essay. Early versions of this work were presented to the 2019 Texas Ecocritics Network workshop and to the 2020 Modern Language Association annual conference, where I benefited from conversation with participants. For their feedback on this essay, as well as their continued support, I am particularly grateful to Paul W. Burch, Brooke Clark, Sonia Del Hierro, Meredith McCullough, and Sarah Jordan Stout.

1   Hetherington (2019: 6) builds from the foundational work of both Geoffrey Bowker (1994) and Star (1999). My use of inversion references Bowker's conception of infrastructural inversion and relies on Star's analysis of how infrastructure is relational and understood perhaps most intimately when it is broken.

2   Much contemporary work sheds light on the detrimental effects of rising seas due to climate change. For example, Orrin H. Pilkey and Keith C. Pilkey's *Sea Level Rise: A Slow Tsunami on America's Shores* (2019) takes up Rob Nixon's (2011) call to attend to the slow violence of sea level rise. I take here a different tact in order to attend to what Pilkey and Pilkey (2019: xv) acknowledge is, and continues to be, an uneven distribution of coastal risk hazards.

3   This information comes from Rebecca Solnit and Rebecca Snedeker (2013: 3). The United States Geological Survey (USGS n.d.) groundwater information reports that land subsidence is an often-overlooked environmental consequence of our land- and water-use practices.

4   For just one recent account of how Mississippi River engineering amplifies the effects of flood hazard in combination with climate variability, see Samuel E. Munoz et al. 2018.

5   Feminist science and technology studies scholars point out how, as Londa Schiebinger (1999: 147), following Susan Merrill Squier, puts it, "metaphors are not innocent literary devices used to spice up texts. Analogies and metaphors . . . function to construct as well as describe—they have both a hypothesis-creating and a proof-making function." For Squier's recent discussion of this, see *Liminal Lives* (2004: esp. 38).

6   I look also to Tiffany Lethabo King 2019 as a model of using figure (the shoal) as metaphor, method of reading, and mode of theorizing the intersections of fields. Jordan Peele's (2017) "sunken place" is not unrelated here as it develops a vocabulary for the ongoing oppression and dispossession of Black people that is seemingly submerged below the surface of twenty-first-century postracial imaginings.

7   Nicole Seymour (2013: 27) observes the "disingenuousness" of referring to Katrina as a "'natural disaster' when in fact most of its greatest harms resulted from social inequalities," and Sharpe (2016: 79) emphasizes the dehumanizing racial logics that conflate Blackness with death such that "Black women and children continue to be cast as less-than-human *victims* and *agents* of 'natural' disasters." Wai Chee Dimock's (2009: 144) analysis of the disaster notes that to see the impact of Hurricane Katrina "as solely a problem of the levees is already to predetermine the solution, making Katrina an event . . . that can be fixed without changing our basic sense of what the sovereign nation amounts to, what it is equipped (or not equipped) to do, and the extent of protection it is able to offer its citizens." Henry A. Giroux (2006: 24) also discusses at length the biopolitics

of disposability, the "murky quicksand" beneath political rhetoric, revealed in the event of Hurricane Katrina.

8   For one account of the contemporary debate over whether literature can adequately represent or respond to climate change, see Stef Craps and Rick Crownshaw (2018: 4) who "hold[] faith with the capacities of the novel."

9   I am encouraged here by Eva Giraud et al. (2018: 64) who call for "engage[ment] with environmental figures to move beyond the narration of entangled worlds," to ask about "the kinds of obligations that particular figures pose and how interventions can be conceived or enabled."

10   For more on this physiography, see *Encyclopedia Britannica Online*, Academic ed., s.v. "Mississippi River," https://www.britannica.com/place/Mississippi-River (accessed September 1, 2019).

11   There is much ongoing research in this area. For just three examples, see Bianchi 2016: 94–95; Jones et al. 2016; and Yu and Michael 2019.

12   Andrea Ballestero (2019: 27), writing of aquifers, argues for the development of a "conceptual vision" that is able to attend to the verticality, in addition to the horizontality, of ecological infrastructure. Ballestero proposes a "circular awareness" that "invites us to understand a sense of being enfolded by our surroundings in a space that is not empty, but dense and potentially pushing against our skin" (27). This perspective follows Stephen Graham's (2004: 21) (after Eyal Weizman 2002) call for the need to develop a "vertical geopolitical imagination" that can attend to the ways colonial powers exploit and extract "key underground resources . . . to fuel the ecological demands of Western urban complexes."

13   I use topological here to distinguish textual spaces from geographical ones and to highlight the figurative abilities of the literary form, which can co-construct, with the reader, the narrative space. For more particular use of the term topology in narrative design, see David Herman 2002: esp. 280–81. Erin James (2015: 56) uses Herman's explanation of topological representation when discussing the shift I point out here, "from a top-down to an on-the-ground representation of space" that can "foster[] intimacy with a local understanding of place."

14   This situated focus borrows from critical regionalist methodology, which, as Krista Comer (2015: 156) describes it, "thinks about issues of place, bodies in place, and knowledges derived not only via textuality and discourse, but from place as a critical location, an orientation, and a material structure."

15   Here I paraphrase Haraway's (2016) call to stay with the trouble. Similarly, LeMenager (2017: 225) focuses on the "everyday Anthropocene" as a way to think against epoch-scale forgetting, and instead think at the "granular and personal" level of the body. I use "intimate" here following Jen Jack Gieseking (2018) and Geraldine Pratt and Victoria Rosner (2012), who argue for a scalar framing of global-intimate.

16   Borrowing from Eve Kosofsky Sedgwick (2003: 131), I suggest here that the focus on rising seas has become a particular paranoia of climate change

discourse, "blotting out any sense of the possibility of alternative ways of understanding *or* things to understand." As Adam Trexler (2015: 82) notes, "Over the last forty years, the dominant literary strategy for locating climate change has been the flood."

References

Ballestero, Andrea. 2019. "The Underground as Infrastructure? Water, Figure/Ground Reversals, and Dissolution in Sardinal." In *Infrastructure, Environment, and Life in the Anthropocene*, edited by Kregg Hetherington, 17–44. Durham, NC: Duke Univ. Press.

Bianchi, Thomas S. 2016. "Effects of Sea-Level Rise and Subsidence on Deltas." In *Deltas and Humans: A Long Relationship Now Threatened by Global Change*, 91–111. New York: Oxford Univ. Press.

Bowker, Geoffrey. 1994. "Information Mythology and Infrastructure." In *Infrastructure Acumen: The Understanding and Use of Knowledge in Modern Business*, edited by Lisa Bud-Frierman, 231–47. London: Routledge.

Britton-Purdy, Jedediah. 2018. "The World We've Built." *Dissent*, July 3. www .dissentmagazine.org/online_articles/world-we-built-sovereign-nature -infrastructure-leviathan.

Comer, Krista. 2015. "Place and Worlding: Feminist States of Critical Regionalism." In *Transcontinental Reflections on the American West: Words, Images, Sounds beyond Borders*, edited by Ángel Chaparro Sainz and Amaia Ibarraran Bigalondo, 153–71. London: Portal Editions.

Craps, Stef, and Rick Crownshaw. 2018. "Introduction: The Rising Tide of Climate Change Fiction." *Studies in the Novel* 50, no. 1: 1–8.

Crownshaw, Rick. 2016. "Agency and Environment in the Work of Jesmyn Ward: Response to Anna Hartnell, 'When Cars Become Churches.'" *Journal of American Studies* 50, no. 1: 225–30.

Dimock, Wai Chee. 2009. "World History According to Katrina." In *States of Emergency: The Object of American Studies*, edited by Russ Castronovo and Susan Gillman, 143–61. Chapel Hill: Univ. of North Carolina Press.

Ericson, Jason P., Charles J. Vörösmarty, S. Lawrence Dingman, Larry G. Ward, and Michel Meybeck. 2006. "Effective Sea-Level Rise and Deltas: Causes of Change and Human Dimension Implications." *Global and Planetary Change* 50, no. 1–2: 63–82.

Gieseking, Jen Jack. 2018. "Size Matters to Lesbians, Too: Queer-Feminist Interventions into the Scale of Big Data." *Professional Geographer* 70, no. 1: 150–56.

Giraud, Eva, Greg Hollin, Tracey Potts, and Isla Forsyth. 2018. "A Feminist Menagerie." *Feminist Review* 118, no. 1: 61–79.

Giroux, Henry A. 2006. *Stormy Weather: Katrina and the Politics of Disposability*. Boulder, CO: Paradigm Publishers.

Graham, Stephen. 2004. "Vertical Geopolitics: Baghdad and After." *Antipode* 36, no. 1: 12–23.

Haraway, Donna J. 1997. *Modest_Witness@Second_Millenium.FemaleMan©_ Meets_OncoMouse™: Feminism and Technoscience*. New York: Routledge.

Haraway, Donna J. 2016. *Staying with the Trouble: Making Kin in the Chthulucene*. Durham, NC: Duke Univ. Press.

Herman, David. 2002. *Story Logic: Problems and Possibilities of Narrative*. Lincoln: Univ. of Nebraska Press.

Hetherington, Kregg. 2019. "Keywords for the Anthropocene." Introduction to *Infrastructure, Environment, and Life in the Anthropocene*, edited by Kregg Hetherington, 1–13. Durham, NC: Duke Univ. Press.

Jacobsen, Rowan. 2011. *Shadows on the Gulf: A Journey through Our Last Great Wetland*. New York: Bloomsbury.

James, Erin. 2015. *The Storyworld Accord: Econarratology and Postcolonial Narratives*. Lincoln: Univ. of Nebraska Press.

Jones, Cathleen. E., Karen An, Ronald G. Blom, Joshua D. Kent, Erik R. Ivins, and David Bekaert. 2016. "Anthropogenic and Geologic Influences on Subsidence in the Vicinity of New Orleans, Louisiana." *Journal of Geophysical Research: Solid Earth* 121, no. 5: 3867–87.

Juhasz, Antonia. 2013. "When They Set the Sea on Fire." In Solnit and Snedeker 2013: 50–54.

King, Tiffany Lethabo. 2019. *The Black Shoals: Offshore Formations of Black and Native Studies*. Durham, NC: Duke Univ. Press.

LeMenager, Stephanie. 2014. *Living Oil: Petroleum Culture in the American Century*. New York: Oxford Univ. Press.

LeMenager, Stephanie. 2017. "Climate Change and the Struggle for Genre." In *Anthropocene Reading: Literary History in Geologic Times*, edited by Tobias Menely and Jesse Oak Taylor, 220–38. University Park: Pennsylvania State Univ. Press.

Madera, Judith. 2015. *Black Atlas: Geography and Flow in Nineteenth-Century African American Literature*. Durham, NC: Duke Univ. Press.

Munoz, Samuel E., Liviu Giosan, Matthew D. Therrell, Jonathan W. F. Remo, Zhixiong Shen, Richard M. Sullivan, Charlotte Wiman, Michelle O'Donnell, and Jeffrey P. Donnelly. 2018. "Climate Control of Mississippi River Flood Hazard Amplified by River Engineering." *Nature* 556, no. 7699: 95–98. doi.org/10.1038/nature26145.

Nixon, Rob. 2011. *Slow Violence and the Environmentalism of the Poor*. Cambridge, MA: Harvard Univ. Press.

NOAA (National Oceanic and Atmospheric Association). n.d. "Hurricanes in History." *National Hurricane Center and Central Pacific Hurricane Center*, USA.gov. www.nhc.noaa.gov/outreach/history/#katrina (accessed September 1, 2019).

Peele, Jordan. 2017. *Get Out*. Blumhouse Productions and Universal Pictures. DVD. Universal City, CA: Universal Studios Home Entertainment.

Pilkey, Orrin H. and Keith C. Pilkey. 2019. *Sea Level Rise: A Slow Tsunami on America's Shores*. Durham, NC: Duke Univ. Press.

Pratt, Geraldine and Victoria Rosner. 2012. "The Global and the Intimate." Introduction to *The Global and the Intimate: Feminism in Our Time*, edited by Geraldine Pratt and Victoria Rosner, 1–27. New York: Columbia Univ. Press.

Schiebinger, Londa. 1999. *Has Feminism Changed Science?* Cambridge, MA: Harvard Univ. Press.

Sedgwick, Eve Kosofsky. 2003. "Paranoid Reading and Reparative Reading, or, You're So Paranoid, You Probably Think This Essay Is about You." In *Touching Feeling: Affect, Pedagogy, Performativity*, 123–52. Durham, NC: Duke Univ. Press.

Seymour, Nicole. 2013. *Strange Natures: Futurity, Empathy, and the Queer Ecological Imagination*. Urbana: Univ. of Illinois Press.

Sharpe, Christina. 2016. *In the Wake: On Blackness and Being*. Durham, NC: Duke Univ. Press.

Solnit, Rebecca and Rebecca Snedeker. 2013. *Unfathomable City: A New Orleans Atlas*. Berkeley: Univ. of California Press.

Squier, Susan Merrill. 2004. *Liminal Lives: Imagining the Human at the Frontiers of Biomedicine*. Durham, NC: Duke Univ. Press.

Star, Susan Leigh. 1999. "The Ethnography of Infrastructure." *American Behavioral Scientist* 43, no. 3: 377–91.

Tatum, Stephen. 2007. "Spectrality and the Postregional Interface." In *Postwestern Cultures: Literature, Theory, Space*, edited by Susan Kollin, 3–29. Lincoln: Univ. of Nebraska Press.

Trexler, Adam. 2015. *Anthropocene Fictions: The Novel in a Time of Climate Change*. Charlottesville: Univ. of Virginia Press.

USGS (United States Geological Survey). n.d. "Land Subsidence." *USGS Groundwater Information*, USA.gov. water.usgs.gov/ogw/subsidence.html (accessed September 1, 2019).

Ward, Jesmyn. 2008. *Where the Line Bleeds*. New York: Scribner.

Ward, Jesmyn. 2011. *Salvage the Bones*. New York: Bloomsbury.

Ward, Jesmyn. 2017. *Sing, Unburied, Sing*. New York: Scribner.

Weizman, Eyal. 2002. "The Politics of Verticality." *Open Democracy*, April 23. www.opendemocracy.net/en/article_801jsp/.

Yaeger, Patricia. 2007. "Introduction: Dreaming of Infrastructure." In "Cities," edited by Patricia Yaeger. Special issue, *PMLA* 122, no. 1: 9–26.

Yu, X., and Holly A. Michael. 2019. "Offshore Pumping Impacts Onshore Groundwater Resources and Land Subsidence." *Geophysical Research* 46, no. 5: 2553–62.

Michelle N. Huang   Racial Disintegration: Biomedical Futurity at the Environmental Limit

Abstract Illuminating how biomedical capital invests in white and Asian American populations while divesting from Black surplus populations, this article proposes recent Asian American dystopian fiction provides a case study for analyzing futurities where healthcare infrastructures intensify racial inequality under terms that do not include race at all. Through a reading of Chang-rae Lee's *On Such a Full Sea* (2014) and other texts, the article develops the term *studious deracination* to refer to a narrative strategy defined by an evacuated racial consciousness that is used to ironize assumptions of white universalism and uncritical postracialism. Studious deracination challenges medical discourse's "color-blind" approach to healthcare and enables a reconsideration of comparative racialization in a moment of accelerating social disintegration and blasted landscapes. Indeed, while precision medicine promises to replace race with genomics, Asian American literature is key to showing how this "postracial" promise depends on framing racial inequality as a symptom, rather than an underlying etiology, of infrastructures of public health.

Keywords Asian American literature, dystopia, speculative fiction, health, comparative race theory

> After all is said and done build a new route to China if they'll
> have you
> Who will survive in America?
> —Gil Scott-Heron, "Comment #1" (1970)

On September 21, 2016, Mark Zuckerberg and Priscilla Chan announced that they were donating 3 billion dollars to their own philanthropic foundation, the Chan Zuckerberg Initiative (CZI), to "cure all diseases in our children's lifetime" (Brink 2016). Earlier the same year, Flint, Michigan, had belatedly declared a state of emergency due to the lead-filled water that had been poisoning its predominantly Black population for almost two years. The two events

American Literature, Volume 93, Number 3, September 2021
DOI 10.1215/00029831-9361293   © 2021 by Duke University Press

seemed to emanate from different worlds—one leaking from the old ruins of a Midwestern postindustrial city plagued by crumbling infrastructure, the other issued forth on the lofty winds of Silicon Valley's technoscientific promise; one postapocalyptic, one futuristic; one a problem, one a solution. The temporal proximity yet unfathomable distance between these white, Asian American, and Black bodies invites several questions: Are Flint's children, especially vulnerable to the deleterious effects of lead on neuropsychological development, included in the CZI's vision? Does toxic exposure count as disease? Is biomedical research the most efficacious investment for curbing death and illness caused by noxious built environments? Can technoscientific solutions ever intelligently contemplate systemic racism?

Flint is not an outlier: proximity to disaster, write the editors of *ASAP/Journal*'s "Apocalypse" special issue, "depend[s] as much on a privileged access to resources as on geographical location" (Hurley and Sinykin 2018: 453). Across the United States, Black, Latinx, and Indigenous people are subject to higher levels of toxic exposure and more likely to live by industrial runoff, hazardous waste sites, air pollution, power plants, landfills, and highways. These damaging externalities contribute to underlying conditions, that medical term of art referring to long-term or chronic illnesses that weaken one's immune system. Black people "experience a higher incidence of hypertension, diabetes, colorectal cancer, infant mortality, and HIV infection rates than any other racial or ethnic group in the country," Latinx people "experience higher rates of diabetes, HIV infection, tuberculosis, cervical cancer, stomach cancer, liver cancer, and liver disease," and Indigenous people experience an "age-adjusted death rate [that] far exceeds the general population, by almost 40 percent" (Ehlers and Hinkson 2017: viii). Underlying conditions, then, extend beyond the individual racialized body and should refer also to the structural and material conditions of damage to which it is subject.[1] This reconfiguration also means that racism—as Ruth Wilson Gilmore (2007: 28) defines it, "the state-sanctioned or extralegal production and exploitation of group-differentiated vulnerability to premature death"—resides not in bodies but rather inheres in the infrastructures of public health that shape natural and built environments.

The disintegration of Flint's lead pipes is a symptom arising from the underlying condition of racism in America. Their disrepair manifests the three dimensions of infrastructure outlined by Akhil Gupta (2018: 65): "a channel that enables communication, travel, and the transportation of goods; a biopolitical project to maximize the health and welfare of the population at the same time as subjecting it to control and discipline; and its role as a symbol of the future being brought into fruition." This means that the ruination of the environs where minoritized people live is both a biopolitical and a futural project.

To illuminate how biomedical capital invests in white and Asian American populations while divesting from a Black surplus population,[2] this article sketches how recent Asian American dystopian fiction provides a crucial case study through which to analyze futurities where healthcare infrastructures intensify racial inequality under terms that do not include race at all. I focus on Chang-rae Lee's *On Such a Full Sea* (2014) but also interweave short analyses of Ling Ma's *Severance* (2018), Rachel Heng's *Suicide Club* (2018), and Ted Chiang's "It's 2059, and the Rich Kids Are Still Winning" (2019). Written by highly professionalized, cosmopolitan, elite, post-1965 East Asian Americans skeptical of technoscientific solutions to an ailing society, these texts critique racial inequality from a position of privilege. They are deeply ironic—as Allison Carruth (2018: 119) writes in her analysis of *On Such a Full Sea*, the novel parodies its apparent genre "via formal disavowals of dystopia"—and tonally flat. Ursula Heise (2015) panned *On Such a Full Sea* (alongside Margaret Atwood's *MaddAddam* trilogy and other texts), castigating the "familiar and comfortable" nature of the novel, whose "focus on details of everyday life make survivalists hard to tell apart from hipsters . . . [and whose] visions of the future serve mostly to reconfirm well-established views of the present."[3]

But what is "the present"? The uneven distribution of the apocalypse is also temporal: Silicon Valley preppers whose analyses of calculated risk compel them to get LASIK and buy land in New Zealand still live in preapocalyptic times, while survivors of transatlantic slavery, Indigenous genocide, and nuclear and chemical warfare live in the postapocalyptic.[4] These different temporalities are displayed starkly in *On Such a Full Sea*, which presents an environmentally degraded world where metropolises are populated predominantly by white and rich, educated East Asian American peoples and ruinously expensive medical care is necessary to survive. Funding for public health languishes while investment in "precision" or "personalized" medicine—meaning patient-tailored preventative, diagnostic, and therapeutic medical interventions—continues to boom. The development of precision medicine (in the novel, "geno-chemo") only benefits this moneyed elite, customizing chemotherapy with genomic data and condemning everyone else to slow death via the toxic exposure that causes the ubiquitous "C-illness."

This emphasis on individualized health solutions, which promise a better future for all, often overlooks structural factors that disproportionately affect minoritized populations. Yet with little explicit mention of race, *On Such a Full Sea* is *studiously deracinated,* a term I develop to refer to a narrative strategy defined by an evacuated racial consciousness that ironizes assumptions of white universalism and uncritical postracialism. While deracination typically connotes uprootedness and alienation, particularly relevant to histories of American enslavement and immigration, I use it here in a more literal sense to denote the erasure of race itself. Counterintuitively, studious deracination challenges biomedical discourse's putatively "color-blind" approach to healthcare, showing how, when aided by environmental deregulation, it sanctions—rather than curbs—structural racism. While infrastructure is often envisioned as synonymous with the notion of public works, as revealed by the gap in provisioning between Flint and the CZI, the instance of healthcare shows "that conceptual alignment is not a guarantee; it is a constant struggle" (Rubenstein, Robbins, and Beal 2015: 577). In sum, as infrastructures of public health (comprising access to a healthy environment and resources as well as to healthcare) deteriorate, the diversion of investment toward advances in individualized biomedical technology amplifies racial health inequity.

The racial logics of the future worlds I consider here are defined less by individual embodiment and more by access to medical treatment; these fictions require a reconsideration of comparative racialization in a moment of accelerating social disintegration and blasted landscapes. Biocapitalism and racism both thrive on the production of environments so bad only few can afford to *relatively* survive them. Specifically, *On Such a Full Sea* shows the need to historicize infrastructures of healthcare by connecting the "history of racial disenfranchisement" to "effects on individuals' present health states" (Bridges 2017: 177). Further, in a book with almost no explicit engagement with contemporary racial discourse, the societal mania for bodily purity—to be "C-free"—can be read as a nefarious allegory of the quest for postracial notions of healthiness, ones that disregard how biopower has always utilized infrastructures of public health to split "healthy" from "unhealthy" populations in order to regulate raced, gendered, disabled, and sexualized bodies. While biomedical developments such as precision medicine promise to replace race with genomics for the benefit of all, *On Such a Full Sea* and other recent Asian American dystopian fiction show how this "postracial" promise depends on framing racism as a symptom, rather than an underlying etiology, of inequitable healthcare access.

We the People

When Barack Obama penned a July 6, 2016, op-ed for the *Boston Globe* elaborating his support for the Precision Medicine Initiative, he did not know the next president of the United States would be Donald Trump. Nor did he know Republican lawmakers were less than a year away from proposing the American Health Care Act, the disastrous healthcare bill that barely failed to pass Senate ratification. The op-ed introduced the Precision Medicine Initiative (since renamed All of Us), which garnered 130 million dollars of key funding under Obama's 2016 budget to support the tailoring of medical treatment to individual patient characteristics, mostly genetic. Channeling an undifferentiated "we" that is difficult to access post-Trump, Obama (2016) writes, "And because precision medicine empowers people to monitor and take a more active role in maintaining their health, it can preempt the hurt and heartbreak many of us have endured when we've seen our loved ones suffer." Obama's promissory missive, which locates suffering at the level of the family and positions healthiness as the responsibility of the individual, skirts both the troubled history of race and genomics as well as the limits of precision medicine to date.

While Obama's postracial "we" in part marks the royal "we" (*pluralis majestatis*) of a single state official, he also invokes the author's "we," in this case the "we" of the American citizenry (*pluralis modestiae*). Obama's nosism both does and does not mirror that of *On Such a Full Sea*, whose most defining formal innovation is the hive mind–like "we" narrator of B-Mor, a fish and vegetable production facility located at the place *we* currently call Baltimore. The novel's "we" is an anonymous someone subject to the "optimizing metrics" that define B-Mor (Lee 2014a: 256), but who discloses little individual history or particularity. Initially, our narrator seems to confirm uncomfortable stereotypes about massed and undifferentiated Asian collectivity. Yet Kandice Chuh (2013) writes that impersonality can be an aesthetic strategy that "de-narrativizes virtuous personhood. It absents the self; there is no hero, no story, no salvation." In other words, the peaceable "we" narrator is not an individual who has yet to self-actualize, but rather a pointed intervention into the desire for a singular speaker. The narrator themself recounts in an oft-quoted passage,

> More and more, we can see that the question is not whether we are 'individuals.' We can't help but be, this has been proved, case by case. We are not drones or robots and never will be. The question, then, is whether being an 'individual' makes a difference anymore. That it can matter at all. And if not, whether we in fact care. (Lee 2014a: 3)

The narrator's disavowal of individuality confounds the expectation that a central protagonist creates action that drives a narrative. Being an individual cannot "make a difference" or solve the problems of systemic racism despite ostensible legal protection ("proved, case by case"). The novel itself serves as a test case for this proposition about individuals making a difference. The picaresque plot follows Fan, a tiny, fifteen-year-old aquaculture diver from B-Mor, as she searches for her boyfriend Reg, a mixed Black Asian teen (and father of Fan's unborn child) who has mysteriously disappeared. Her inaction and "low animacy" are constantly evinced by comparisons to nonhuman life such as her fish or "an arbitrary plant or small tree in a section of counties bush" (Fan 2017: 685).[5]

Agency is instead shown to be nefariously distributed across governing infrastructures. Though no laws exist to explicitly enforce racial segregation in this future United States, race is nonetheless established through three socially stratified spaces—the Charters, the facilities, and the counties—demonstrating Adrienne Brown's (2017: 27) observation that race is "always shaped in some way by the built environment." These different areas are physically demarcated by gates and checkpoints, with "secured, fenced tollways" that only allow Charter cars or other authorized vehicles (Lee 2014a: 34–35). Spatial organization also determines biopolitical citizenship: while the Charters effectively function as a "conservative one-percenter's free market fantasy . . . [where] both income and property taxes are kept at 'negligible levels,' and virtually all goods and services, not only those pertaining to health and medicine, are privatized" (Barrish 2018: 302), the facilities have a limited system with caps on expenditures and life-saving measures, and the counties languish without formal insurance or medical care. In general, upper-class white and Asian American professionals populate the Charters, descendants of recent Chinese immigrants reside in the facilities, and everyone else, including Black people (referred to within the novel as "natives"), has been displaced to the counties. (There are notable exceptions, including mixed-race B-Mors as well as Latinx and Southeast Asian care workers in the Charters.) B-Mor's New Chinese denizens constitute the working class whose labor sustains the elite professional class of the Charters, while the counties exist as a no-man's-land with no discernable governance or rule of law. In addition to a conspicuous absence of the notion of the panethnic political entity "Asian Americans," there is only fleeting mention of chattel slavery, Native American genocide, and Latinx migration.[6]

On Such a Full Sea takes place in a globalized context. In an inversion of yellow peril discourse that associates Asians and Asian Americans with disease and pestilence, it is American capitalism that pollutes Asia, specifically China.[7] The original B-Mors are environmental refugees. The narrator reflects on their former home of Xixu City:

> that someplace, it turns out, is gone. You can search it, you can find pix or vids that show what the place last looked like, in our case a gravel-colored town of stoop-shouldered buildings on a riverbank in China, shorn hills in the distance. Rooftops a mess of wires and junk. The river tea-still, a swath of black. And blunting it all is a haze that you can almost smell, a smell, you think, you don't want to breathe in. (Lee 2014a: 1)

The toxic atmosphere of New China forces their immigration to the United States, the sullied river signposting a poisoned and stagnant future. Lee's B-Mor was inspired in part by a 2011 visit to the special economic zone of Shenzhen (Lee 2014b); this revelation sets up a parallel between postindustrial American cities and global production hubs as sites of American capitalism's ruination.

The city of Shenzhen appears even more explicitly in Ma's *Severance*, where the Asian American narrator, Candace Chen, works for a New York publishing company that outsources its production there. The workers at the Shenzhen factory are being poisoned by the dust created by the manufacturing process of the keepsake gemstones that accompany the Gemstone Bible Candace is overseeing: "Pneumoconiosis is an umbrella term for a group of lung diseases . . . they've been breathing in this dust and developing lung diseases, without their knowledge, for months, even years" (Ma 2018: 24). But to Candace's supervisor, the poisoned workers are a production setback, not a health issue. Candace is thus not a representative of Asian American "identity" so much as she is a corporate representative of how Asian American professionals support American multinational corporations in the production of toxic environments for subaltern workers—in this case, other Asians. Both texts query the relationship between post-1965 East Asian Americans—largely educated and upwardly mobile—and the Asian workers whose labor helps underwrite the global economy.

As a result of the ubiquitous pollution suffusing their world, everyone in *On Such a Full Sea* inevitably dies of "C-illness," whose only "treatment [is] sold at hourly auctions" that even Charter physicians cannot always afford (Lee 2014a: 273). The elided "C" "strongly implies 'cancer' while also inviting interpretations like 'crash,' 'climate,' 'capital,' and 'China'" (Fan 2017: 680–81). The narrow fixation on C refracts how broad focus on a universal "cure" diverts money and attention from the underlying and durable systems of inequity that adjudicate healthiness. When untethered from social context, the discourse of improving individual well-being opportunizes biomedical capitalism. Akin to the disease-free future offered up by the CZI, the fantasy of being "C-free" allows moneymaking to proceed unabated in the name of a greater good that can only be accessed by an increasingly narrow sliver of the world's population.

Thus, the irony of Reg being "cellularly pure," which is speculated to be the cause of his mysterious disappearance, is the exception that proves the racialized rules of C-death and vulnerable populations. Reg is mixed Black Asian, with "Afro-type hair" inherited from an ancestor who was a "native" girl (Lee 2014a: 66).[8] The will to knowledge about his C-immunity is very strong: what has exempted him? The narrator advances several possibilities, from his "particular fusion of original and native blood" to his idiosyncratic distaste for fish (102), thematizing pseudoscientific belief in the hybrid vigor of mixed-race people as well as in fad diets like clean eating. Yet the hope that a mixed-race individual could provide a universal cure from within his biological body proffers a cautionary tale about the promise of any panacea that distracts from broader infrastructures of public health, especially ones where Blackness is only visible as mixedness.

For within the world of *On Such a Full Sea*, the cause of the C-illness has not evolved over time; what *has* changed are technologies of detection and who can access treatment. Rather than healthy being the baseline and sick being the exception, the constant testing has created a wellness culture where everyone is constantly ill and in anxious need of treatment. Graphs of mortality rates over the last one hundred twenty-five years show that

> though it appeared there was vast improvement after controlling for nascent-stage diagnoses, which is how Charter survival rates were measured, Charters didn't actually live more than a few years longer than they did back then. People now just knew much earlier that they were diseased, literally sometimes mere days into the condition. And while they were being 'cured' with all the therapies available now, it could be argued that they were never actually 'well,' given the constant stress of regimens and associated side effects. (277)

The extended duration of C-illness does little but uncover further potential for capitalization, as Charters are not living appreciably longer; the "vast improvement" is artificially generated by earlier start dates. The passage reflects Jasbir K. Puar's (2009: 166) formulation of "prognosis time," which she develops to "put pressure on the assumption of an expected life span—a barometer of one's modernity—and the privilege one has or does not have to presume what one's life span will be, hence troubling any common view of life phases, generational time, and longevity." The diminishing returns on investment within the novel paint a damning portrait of infrastructures of health that reward the hoarding of both public concern and the resources necessary to administer these increasingly individualized treatments. Given the well-known and widespread effects of toxic environments, the unrelenting destruction of the world that future generations will inherit can only be convincingly explained by relative status anxiety. The goal is not to produce an average longer life span but to widen the gap between access and inaccess, possession and dispossession. In truth, the myth of being "disease free" depends on a fantasy of bodily purity that has never and will never exist, and whose logical extension is immortality, whereby "the wealthy can purchase the fantasy of a regenerative body at the expense of the health of other, less valuable bodies" (Waldby and Mitchell 2006: 187). Such racialized moral hazard is not contemplated in more utopic accounts of technoscientific bodily development.

To generate these varying prognostic timelines, everyone in the Charters and the facilities is subject to biomonitoring that predicts their mortality. These mandatory tests include an "annual blood panel" where "protein, sugar, fat, hormone, vitamin, and numerous other levels [are] collected and tabulated," as well as C-testing: "Eventually, everyone will express it, the blood panels show this . . . our tainted world looms within us, every one" (Lee 2014a: 65). All the listed biomarkers confirm the idea that one's life expectancy is comprised of discrete individual metrics, rather than proximity to pollution or access to healthcare. Illness is supposed to be the great leveler—as the narrator reminds us, "Nobody goes C-free—nobody" (101)—but in fact it indexes racial inequality. B-Mors are duly included in the population to be studied but not in the "we" of the population deserving treatment. Finally, what *causes* C cannot be measured by all the biometric data available, much less by the overdetermined biological marker of "blood"—it is the asymmetric burden of "our tainted world."

Studious Deracination, or Whither Race?

As a text published in the immediate wake of Black Lives Matter (BLM) that eschews contemporary terminology regarding minority identity, *On Such a Full Sea* initially reads as a perniciously postracial novel, bolstering claims about the proximal whiteness of Asian Americans. But I would argue the novel is *studiously deracinated*, a term I develop to extend Ramón Saldívar's (2013: 14) formulation of "postracial aesthetics," which "define the historical contradictions in the justification of racial injustice, discrimination, and oppression in terms that can then be related to the form and language of the literary text." By studiously deracinated, I mean a stylistic strategy defined by an evacuated racial consciousness, often accompanied by an affective flatness, that nonetheless ironizes assumptions of white universalism and uncritical postracialism. *On Such a Full Sea* and other contemporary Asian American dystopias reveal how deracination works hand in hand with racial inequality by requiring interpretative practices sharply sensitive to what Eduardo Bonilla-Silva (2017: 2) has termed "color-blind racism," which "explains contemporary racial inequality as the outcome of nonracial dynamics."

The novel predates the 2015 murder of Freddie Gray by Baltimore police, but it is informed by both BLM and the underlying conditions laid decades prior.[9] Its studious deracination thus presents a particular critical challenge to post-1965 Asian Americans; while the novel might seem to invite a model minority reading practice "that exemplifies neoliberal finance capital's disavowal of colonialism and racism as relics of the past" (Hong 2018: 112), it also makes available a damning critique. In *On Such a Full Sea*, there are *ethnic* markers — for example, B-Mor's primary industries are the factory farming of fish and vegetables, an echo of Filipino Alaskeros, Korean *haenyeo* (women divers), and the Japanese American farmers of California. Conspicuously missing, though, is a longer sense of Asian American history, an absence that reinforces the assimilative trope of Asian Americans playing the foreigner over and over again.

The doubled reading practice necessary to excavate buried racial references, histories, and forms contributes to what Christina Sharpe (2016: 13) calls "wake work," which is invested in "plotting, mapping, and collecting the archives of the everyday of Black immanent and imminent death, and in tracking the ways we resist, rupture, and disrupt that immanence and imminence aesthetically and materially." Careful appraisal of the new history offered in *On Such a Full Sea* reveals that the Asian immigrants referred to as the "originals" (a generation sometime after our contemporary moment who arrived "nearly a hundred years ago"; Lee 2014a: 13) displaced not just Black people but also their livelihoods. The shadowy "directorate" established the facilities in cities like Baltimore and Detroit where "there were, in fact, numerous existing businesses when the originals arrived, businesses run by the smattering of natives who had stayed on, whose deeds and leases to their properties were unilaterally voided and reassigned" (69).[10] B-Mors live in row houses emptied of Black residents. And the street "Longevity Way," formerly "North Milton" (14), signifies the Black neighborhood Middle East that has been the target of aggressive urban redevelopment since the early aughts (Johns Hopkins University, a school notorious both for its role in medicine and its STEM-focused Asian American population, has been a key contributor).

Studious deracination widens our interpretive purview from people to built environments and infrastructure. Another studiously deracinated site is Seneca, the Charter village where Fan's brother Oliver, a doctor *and* venture capitalist, lives. In the Charters, wealth and class trump race—"there were tallish, attractive people of various races and ethnicities going about" (163). Yet this postracial Seneca refers not only to the displaced Seneca peoples but also to Seneca Village, a free Black community founded in 1825 that, in the 1850s, was demolished to make way for Central Park.[11] (This history of forced removal also colors the derogatory nickname "Parkies" given to the Black residents of B-Mor who take over a city park after their eviction.) Tracing "the earth as a site of black memory," Jennifer C. James (2011: 163) sketches how "ecomelancholia's historical and memorial disposition defends against mourning's call to prematurely forget. It responds to the cumulative losses of nature, land, resources, and to traumas tied to those losses, such as death, deracination, and dispossession: it is activated by ongoing and interrelated social and political violence, including the catastrophes of war, genocide, and poverty" (167). The ecomelancholia of Seneca's name inflects contemporary New York City, whose rapid gentrification has eviscerated the local minority populations.

Unearthing these buried and imbricated histories also requires theories of relative racialization. Claire Jean Kim's (1999: 107) work on racial triangulation is illuminating here: she argues the dominant group of white Americans valorizes the subordinate group of Asian Americans over the subordinate group of Black Americans "in order to dominate both groups, but especially the latter." One of Kim's examples is William Pettersen's 1966 *New York Times Magazine* article "Success Story, Japanese-American Style," which contrasts postinternment Japanese Americans, purportedly able to overcome the hardship of internment, with the "problem minority," i.e. Black Americans (Kim 1999: 119). In racial triangulation, the Asian American wedge function is kept in place by the fantasy of ascension into the higher class; B-Mors are included in minor capacities within Charter economy and culture, "mostly [as] bystanders or else hardworking service people" (42), and a miniscule percentage who "do extremely well on the yearly aptitude Exam" are "Chartered" and jump social spheres (27). This putative meritocracy is even visible in the name B-Mor, which denotes the imperative to keep improving, to be productive, to "be more." When the narrator plaintively asks, "Have we not done the job of becoming our best selves?" (Lee 2014a: 21), "our best selves" invites individual achievement to occlude the structural subjugation of others.

I have been trying to show that reading contemporary Asian American literature chiefly through proximity to whiteness fails to grasp how it fits into a broader field of racial formation. Even as racial consciousness is elided as an explicit discourse, *On Such a Full Sea*'s collectively voiced "we" narrator peaceably acknowledges there *is* racism—specifically manifested in anti-Blackness and unequal distribution of resources—but no framework for identifying or adjudicating it.

> We won't say it or admit that we know, but we can all appreciate how people with some part of Reg's native line can be very subtly or unwittingly lodged in the lee of prime conditions. Everyone knows that certain spots in the tray don't receive quite the same flows of water and air, where perhaps the nutrients are either diluted or oddly concentrated, and the green shoots there might on first glance look fine but are, in fact, just that bit leggier, more prone to blight. (108)

What our narrator conveys to us is that "racism without racists" (Bonilla-Silva 2017) is not so different than racism with racists. Access to resources that allow for flourishing has always been what determines healthiness. The passive voice here importantly marks the diffusion of responsibility, whereby "people . . . can be very subtly or unwittingly lodged" even though "everyone knows." The work of racism, traceable in the sanguine-tinted language of "diluted" and "oddly concentrated," has been offloaded to infrastructures of public health, which deliver difference "subtly" and "unwittingly."

Contra the broad breadth of environmental deprivation, biomedical technologies of individual health only benefit a select, privileged few. In *On Such a Full Sea*, the standard of care for C is "geno-chemo"; the fictional technology's name suggests a further grafting of genetics onto the existing regimen of chemotherapy, i.e., precision medicine. The people who live in the counties have no access to healthcare beyond unregulated emergency clinics, yet geno-chemo, which is engineered only in Charter labs, is presented as everyone's lone hope. The far-reaching scope of C thoroughly eclipses the narrowness of the individualized solution.

Precision medicine's promise of hope for all Americans does not contemplate such scarcity. As defined by the National Research Council (2011: 124), precision medicine refers to

> the tailoring of medical treatment to the individual characteristics of each patient. It does not literally mean the creation of drugs or medical devices that are unique to a patient, but rather the ability to classify individuals into subpopulations that differ in their susceptibility to a particular disease or their response to a specific treatment. Preventive or therapeutic interventions can then be concentrated on those who will benefit, sparing expense and side effects for those who will not.

Slippages abound here between individual and population; the "individual" that precision medicine names is one that is always already captured in a subpopulation, one generated by information technology and algorithms of value. The statistical impulse of precision medicine is normatively constituted, and its fundamental "emphasis . . . on *lowering risk* tends to encourage a far less nuanced—and ultimately paradoxical—desire to approximate an ideal type, namely, an individual with the lowest possible risk for all known conditions" (Mitchell 2016: 382; emphasis original). These calculations only make sense if structural inequality is bracketed from the equation, yet this official definition of precision medicine is studiously free of any reference to race.

Ironically, one of precision medicine's putative promises is postracialism: "Will Precision Medicine Move Us beyond Race?" queries the May 2016 *New England Journal of Medicine* article on the topic. The article's first author, investigator and senior advisor at the National Human Genome Research Institute Vence L. Bonham, states in an NIH press release that "eventually, precision medicine will revolutionize the utility of race in prescribing medications . . . We have highlighted the importance of genomic knowledge and the precision medicine approach in moving us beyond the use of crude racial and ethnic categories in the care of the individual patient" (Mjoseth 2016).[12] Conflating race and genes as medical inputs, the precision medicine approach hinders healthcare equality by ignoring that race is constituted not individually but by "neighborhoods and communities, and larger-scale systems of governance" (Ehlers and Hinkson 2017: xxi). Race is not an imprecise approximation of genes; genes are a grossly inadequate approximation of race, which is contextual, multivalent, and socially distributed.

Furthermore, members of "crude racial and ethnic categories" have always played outsized roles in generating medical discoveries—as research subjects. From vivisections to the speculum, from Tuskegee to the Marshall Islands, medical progress accumulates through experimentation on vulnerable populations in the name of "we."[13] Susette Min (2016: 851) writes, "in contrast to past and recent forms of scientific racism, Reg is seen as a solution rather than as a problem," but in reality, racial bodies have always enabled scientific and medical discoveries—most pertinent to *On Such a Full Sea* is Henrietta Lacks, the thirty-one-year-old Black woman whose "immortal" cervical cancer cells were harvested without her consent at Johns Hopkins in 1951. The first cells able to be kept alive indefinitely, HeLa led to countless scientific discoveries (including the polio and COVID-19 vaccines) that delivered untold millions to biopharma companies and nothing to the Lacks family.

Attempts to genetically engineer equality at the individual level will always fail if structural racism is overlooked. Such is the implication of Ted Chiang's *New York Times* op-ed from the future, "It's 2059, and the Rich Kids Are Still Winning" (2019). The fictional op-ed describes a novum called the "Gene Equality Project," a "philanthropic effort to bring genetic cognitive enhancements to low-income communities" through "modifications to 80 genes associated with intelligence" that together typically raised each child to "an I.Q. of 130" (Chiang 2019). Essentially genetic affirmative action, the Gene Equality Project allows Chiang to explore whether intelligence can be found inside or outside an individual body—if intelligence is innate, the Gene Equality Project should work. But it doesn't—the children who receive the gene modifications as a charitable handout do not succeed at the same rates as their peers whose parents paid for the modification. The one sentence explicitly mentioning race reports that because many of the beneficiaries of the Gene Equality Project are not white, its failure has emboldened white supremacists to affirm racist theories of evolution. Yet the question has never been whether intelligence is hereditary, though what intelligence is and whether IQ tests measure it are far from settled issues. The real problem is that "the Gene Equality Project didn't offer any resources besides better genes, and without these additional resources, the full potential of those genes was never realized." Against the logic of precision medicine, Chiang's short story suggests fixing "broken" individuals does not work because the barriers to success are not individual to begin with. If "it has long been known that a person's ZIP code is an excellent predictor of lifetime income, educational success and health," then overemphasizing genetic notions of intelligence obscures the fact that the only truly comprehensive solutions to racial injustice are ones that address resource asymmetry and inequality. This future world is disturbingly oblivious to low-technology possibilities, such as improving public school funding. Not only are individualized stratagems incapable of resolving structural racism's problems, but in practice, they often exacerbate them.

As Chiang's story suggests, race, "an epistemological category of white supremacy that maintains structures of violence and dispossession" (Rana 2015: 204), cannot be encapsulated inside a single body. Yet claims that biomedical development has rendered race obsolete persist. Rejoining Bill Clinton's beatific pronouncement upon publication of the Human Genome Project (HGP) data in 2000 that "all human beings, regardless of race, are more than 99.9% the same," Dorothy Roberts quipped in 2011 that "reports of the demise of race as a biological category were premature" (both quoted in Nelson 2016: 13). In the tradition of the one-drop rule that adjudicated Blackness, disproportionate focus on that 0.01 percent continues to dominate genetic research.[14] There is a degree to which Clinton's universalizing statement was and remains true, but the more important finding is that 0.01 percent of genetic variation is not what comprises racial groups, which "are not products of biological or genetic differences in the human population but rather are biologically arbitrary groupings of humans that are formed as a consequence of political, social, and/or economic forces" (Bridges 2017: 157). Indeed, precision medicine is the furthest thing from an evolution of scientific knowledge beyond the "crude" sociopolitical category of race—making race invisible has always been one of racism's most successful formal strategies.

Assimilation as Condition

The new medicine of promise in *On Such a Full Sea* is the experimental C-drug "Asimil," developed by Fan's brother Oliver (formerly Liwei), who was successfully Chartered out of B-Mor as a child. Asimil's name sends up euphonic, consumer-tested pharmaceutical names such as Ambien and Lyrica; it also clearly invokes "assimilation." Assimilation's apotheosis is the model minority, a Cold War construction "whose apparently successful *ethnic* assimilation was a result of stoic patience, political obedience, and self-improvement" (Lee 1999: 145, emphasis original). Oliver himself is a doctor-cum-businessman "in his early thirties, a blood C-specialist at the medical center" who has been developing Asimil "for the last eight years, starting from when he was still a medical student. Although just 60 percent efficacious in year-long trials, the drug had been deemed promising enough that all three major pharmacorps joined in a frenzied auction for his patent" (Lee 2014a: 275). Oliver's position as a successful model minority subject is underscored by his occupation as a hematologist-oncologist, a doctor who specializes in bad blood. Along with his wife Betty, Oliver decides he will not return to medical practice after the Asimil sale. Instead, he will follow Chan and Zuckerberg (and Bill and Melinda Gates) into philanthropy by "running the charitable foundation they were going to start, maybe for the benefit of Charter helpers' or even counties children's health care, though of this they weren't yet sure" (304).

What might the Asimil's 60 percent success rate mean in relation to the post-1965 Asian American model minority's aspirations toward whiteness? The drug's name can also be read, "As I'm ill," signaling the condition of never being well. David Palumbo-Liu (1999: 396) has critiqued the model minority reading practice of Asian American literature that relies tropologically on the "presumption of a particularly constructed ethnic malaise" that is cured by the ethnic subject "coming-into-health" through assimilation. Through Asimil and Oliver, the novel reconfigures assimilation, historically and still commonly viewed as the "cure" to the problem of racial difference, as the condition in which the racialized body can never be well. Following Palumbo-Liu, assimilation is, like the ambition of eradicating all disease, a false promise that allows Asian Americans to labor under its "cruel optimism" (Berlant 2011) and whose racial triangulating function is to buffer white Americans from Black Americans. "Successful assimilation" is a paradox akin to the one of "perfect health," whereby only some Asian Americans, like Oliver, conditionally succeed. By implication, the Black body is figured as beyond assimilation, pathological and incurable. But as studious deracination shows, racism is a white problem that is distributed onto and represented as contained within racialized bodies. *It is a white dis-ease with raced bodies.*

Dismantling assimilation's curative logic requires more than a refusal of the model minority myth—it also requires a rejection of whiteness. Robert E. Park and Ernest W. Burgess (1969: 735), the Chicago School sociologists, defined assimilation as "a process of interpenetration and fusion in which persons and groups acquire the memories, sentiments, and attitudes of other persons or groups, and, by sharing their experience and history, are incorporated with them in a common cultural life." But Asian Americans are already American, so assimilation into the dominant culture really means the reproduction of "whiteness as the national identity" (Park 2015: 14). By dehistoricizing difference as cultural essentialism while leaving the individualistic logic of race intact, Park and Burgess's sociological definition also problematically "reinscribes as fact the fiction of a unitary national culture, ignores the interlinking of race and gender in cultural formation, and valorizes upward economic mobility in a way that accepts liberal capitalism as a politically neutral index of success and failure" (Cutler 2015: 6). If the conscribed success of the Asian American model minority subject is meant to affirm fantasies of American individualism, *On Such a Full Sea* shows structural problems cannot be solved with the mere presence of exceptional postracial individuals—neither the model minority Asian American STEM worker (Oliver) nor the mixed-race subject (Reg).

For racialized people are never racialized only as individuals but also as representatives of populations: at the turn of the twentieth century, yellow peril discourse figured Asian American immigrants as a diseased horde accompanied by vector-carrying agents like rats, pigeons, and cockroaches. Mel Y. Chen (2012: 194) observes that "immune systems are themselves constituted by the intertwinings of scientific, public, and political cultures together. Even further, we know that the medicalized notion of immunity was derived from political brokerages . . . discourses on sickness bleed from medical immunity discourse into nationalist rhetoric." Accordingly, when Oliver is first introduced, his anglicized name and immaculate appearance symbolize his transcendence of his New Chinese origins. He is described as "unmistakably, irrefutably, clean, as though he had showered twice, a third time, then gone back and fine-scrubbed himself again" (Lee

2014a: 282). By contrast, the counties people are described as reeking and fetid, i.e., "the sharp rot of human settlement" (13), evoking Bruce Robbins's (2007: 26) observation that "the smell of infrastructure is the smell of the public."

What if race is not the disease to be eradicated but the very infrastructure that organizes our shared world? In *On Such a Full Sea*, "the all-powerful C-therapy industry" is embodied by the belief that "there [is] always a cure" (Lee 2014a: 277). But for whom? Despite the Charters' exclusive access to staggeringly expensive regimens, the standard geno-chemo treatments have stopped working: "A growing number of patients, after near lifelong serial therapies (some from when they were in preschool), had stopped responding to the treatments altogether" (277). Eradicating C is an ever-receding shoreline, paralleled by the proposal to raise the Exam score cutoff allowing B-Mors to become Chartered from the top 2.0 percent to 1.25 percent (292).

The Charters' obsession with individual wellness and success belies the diseased world—not only the poisoned banks of Xixu City but also the increasingly common "bird and swine flu epidemics" (59) and another, more recent, deadly zoonotic pandemic that resulted in the outlawing of all pets.

> Initially the cats got sick, and then the dogs, followed by the hobby livestock, but then a small percentage of the human population became infected, which wouldn't have been so catastrophic had not nearly all of those unfortunate people died. The affected villages were immediately locked down, Charter epidemiologists flown in from around the world to determine what was causing the illness (expressed in a catastrophic hemorrhagic fever) and how it had crossed multiple species barriers; while they were working, all pets and animals in the affected villages were ordered destroyed, whether sick or not. (116)

It is impossible to contemplate any modern pandemic without situating it within globalized networks of factory farming, international trade, and pollution.[15] All the animals are indiscriminately sacrificed, even though it is humans who created the conditions for disease transmission and spread. A dark parallel is not explicitly drawn, but suggested, between the biopolitical management of the sick animals and the disappeared native Black population of B-Mor. In our current moment of COVID-19, one cannot ignore the links between the trashed animals at meatpacking plants around the United States and skyrocketing rates of infection among the Latinx bodies who work there, cheek by jowl. As Priscilla Wald (2008: 49) writes, "communities are populations conceived, in effect, as immunological ecosystems, interdependent organisms interacting within a closed environment marked by their adjustment to each other's germs." The pandemic has laid bare the woeful state of infrastructures of public health in the United States, where underresourcing, disaster capitalism, and lack of protections for essential workers have had a devastating impact on Black, Latinx, Indigenous, and other communities of color (Flagg et. al. 2020) who absorb the collective risk while the privileged work from home.

The near-future New York City of Rachel Heng's 2018 novel *Suicide Club* works from a by now unsettlingly familiar scenario in which Black people have been wholly displaced, seemingly due to the hyperbolic extension of contemporary gentrification. Lea Kirino, the protagonist, works in "HealthFin" (a portmanteau of *health finance*) as a trader on the organ market, where she represents the "Musk account." The social imperative to healthiness has been formalized as law itself, requiring people to eat "Nutripaks" of balanced meals "optimized for normal human beings" (Heng 2018: 63) and forbidding them from opening windows and exposing themselves to the toxic environment. The affective world is as stifled as the environment, with biometric surveillance even more precise than the health "stats" of *On Such a Full Sea*. People are born with a number that corresponds to their projected life expectancy, and then they labor to maintain or increase it. Early on, an upper-class socialite worries that a scare must have "taken at least three months off my number" (7). Lea is already one hundred years old, a privileged member of the "Second Wave" with a life expectancy of three hundred. As members of the titular "suicide club" demonstrate by self-immolating in viral videos, the only escape from the constant surveillance and monitoring is death.

The novel also skewers contemporary wellness culture's obsession with antioxidants, cortisol levels, and self-care. Patented upgrades available to "Lifers" like Lea—"DiamondSkin™, ToughMusc™, Replacements" (2)—parody contemporary cosmeceutical treatments and manifest Olivia Banner's (2017: 12) observation that "once the body's pieces or data about those pieces have been translated to a state that is useful in another context, those parts are converted into value, a commodity that can be bought and sold."[16] The results of these treatments are macabre: the Lifers' piecemeal replacements result in bodies that are "misaligned," meaning one's skin can rot and peel away even as their top-of-the-line replacement heart keeps beating. Prevailing laws about the sanctity of life demand they be kept alive. Negating that key term for the promise of Civil Rights–era enfranchisement, these unfortunate souls literally disintegrate.

Instead of countenancing the wicked problem of the disintegrated bodies left in its wake, though, this future high society is borne forth on the assurance of a coming "Third Wave"—the promise of immortality. The political valences of this term, which evokes not only a new wave of technology but also of radical politics (e.g., first- and second-wave feminism), cannot be lost on scholars of social movement history. The fantastical "Third Wave" presents a heuristic for recent Asian American dystopian fiction that animates the threat of a final evacuation of racial consciousness and history, illuminating the stakes in reading these worlds—and ours—as postracial rather than ones where race has been violently expunged. The studious deracination of these worlds enjoins readers to attune themselves to the purposeful elision of minority history and comparative racialization, and to stare down the encroaching dystopia of our own postracial present. For post-1965 East Asian Americans in particular, the inevitable advent of a coming "Third Wave" is a warning that we may be in "misalignment" with a more equitable future.

That better future is not foreclosed, but its arrival is far from certain. Unsettling irresolution marks *On Such a Full Sea*'s end. The novel does not conclude satisfactorily—readers have no idea whether or not Fan finds her Reg, who is being studied by the very same pharmacorp purchasing Oliver's Asimil. The final, hurried moments reveal that Oliver, who has noticed Fan's pregnancy, has schemed to hand over his sister to seal the floundering Asimil deal. (The pharmacorp is more interested in a C-free individual than an imperfect, costly treatment.) Asha Nadkarni (2019: 7) reads this deferred conclusion optimistically, arguing that "against eugenic notions of purity, here mixed-race vitality is associated with reproductive futurity and potential salvation." Yet in putting her faith in Reg and Fan's unborn baby, Nadkarni ironically reproduces the same logic of "always a cure" as the C-industry. Rather than hope, I believe the novel ends with a condemnation of the model minority Asian American who betrays his own sister and her mixed Black unborn child to medical experimentation in order to secure his financial fortune. Moreover, Reg himself is already mixed-race, as is the rest of his family, the Xi-Jang clan,[17] who have all tested positive for C—being mixed is no assurance against disease nor racism.

We cannot find Reg because he is the false promise of a cure, and there can be no cure originating from within a raced individual. Like Henrietta Lacks, Reg's Black body is valuable not as life but as tissue. His family members and the rest of the native B-Mors are not so lucky, relegated to conditions of unflourishing either in B-Mor or in the counties. Reg's absence is a question that leads not to some shared universal salvation but to the other Black bodies that lie beneath the repurposed urban infrastructure of B-Mor, and the United States writ large. As studious deracination has shown, racism works outside-in, not inside-out, meaning its remedy will not come from "within" the exceptional individual who has escaped their circumstance.

Infrastructures delineate futures. Their design dictates who can dwell and thrive within them, and as such they are instrumental to "the biopolitical project of creating citizens who share the goal of inhabiting a modern future" (Gupta 2018: 68). To be concerned with the inequity mapped out by these bad Asian American futures, then, requires a deeper understanding of race's imbrications with the putatively colorblind proportioning and distribution of worldly resources that adjudicate healthliness. Put another way, infrastructure is always racialized, but in order to fully grasp the dystopic reality of Flint—as well as places like the San Joaquin Valley and the Marshall Islands—it is also crucial to think of race *as* infrastructure. Understanding race as infrastructure entails rebuffing the amnesiac dissolution of the sunken racialized histories and bodies that built the future we call our present, as well as recognizing and refusing the triangulating, prophylactic function of Asian American identity in a relative field of racial formation between whiteness and Blackness. It requires uncathecting from individual raced characters and studying the infrastructural composition of built and natural environments.

Here, the significance of *On Such a Full Sea*'s "we" narrator comes into fuller focus. Anne Anlin Cheng (2020: 224) writes, "we cannot critique the narrating 'we' in the novel without also implicating ourselves and our own political longings." If the narrative "I" always presupposes the individual as the template for understanding the world, "we" always refracts the tension between individual and population inherent to both biomedical and racial discourse. For racialized individuals, the suggestion that we jettison "I" when we have never been "I" is a bitter pill to swallow, but it is not as futile as the promise of overcoming one's ethnic malaise and then coming-into-health. If, as the narrator initially queried, "the question, then, is whether being an 'individual' makes a difference anymore," the novel offers no person-alized solutions. We might read this as narrative failure or as the pro-ductive frustration of our own readerly desire for a precise, therapeu-tic resolution. The answer to the narrator's question? *It never has.* The fatuous phantasm of individuality has always been a reading practice, against which the novel presents perceiving studious deracination as only the first step toward apprehending underlying conditions of racial dispossession and resistance.

Michelle N. Huang is assistant professor of English and Asian American studies at Northwestern University. "Molecular Race," her current book project, explores post-humanist aesthetics in contemporary Asian American literature. Her other writing on race, science studies, and Asian American literature has appeared or is forthcoming in *Journal of Asian American Studies, Contemporary Literature, Amerasia, Verge: Studies in Global Asias,* and *Post45: Contemporaries.*

Notes

Thank you to Tina Chen, Henry E. Chen, John Alba Cutler, Chris A. Eng, and Justin L. Mann.
1   Environmental racism is not just a function of income: "African Ameri-cans who earn US$50,000–60,000 annually—solidly middle class—are exposed to much higher levels of industrial chemicals, air pollution and poisonous heavy metals, as well as pathogens, than are profoundly poor white people with annual incomes of $10,000" (Washington 2020: 241). As Jennifer C. James (2020: 692) writes, "The vulnerabilities are structural. 'Underlying conditions' and 'comorbidities' are merely other words for anti-Blackness." Sarah Wasserman (2020: 531) uses the term "infrastruc-tural racism" to refer to "the racist social forms embedded in the built environment."
2   In this article, I refer primarily to the racial triangulation between Black and post-1965 East Asian American populations. Many of these readings apply, but with significant differences, with respect to Latinx and Indige-nous racialization, as well as within the panethnic group of "Asian Ameri-can" with regard to Southeast Asian Americans and low-income older Asian American immigrants.
3   Related terms to this affective diffusion are Heather Houser's (2014: 12–13) "ecosickness," or "the interdependence of narrative strategies and affect and experiments with the ethicopolitical effects of emotional idi-oms," and Jessica Hurley's (2020: 9) "nuclear mundane," which "redefine[s] the nuclear object as continuous with a set of militarized infrastructures rather than as their exceptional end point."
4   See Osnos 2017 and Whyte 2017. The fiction I discuss also trails the zeit-geist of the 2008 financial crisis—*Severance* eerily reimagines the 2011 Occupy Wall Street protest in a pandemic.

5 Rachel Lee (2018: 521) writes of "non-heroic agency" and observes how Fan represents "a simple and miniscule part of a vital ecology where the decomposition of one set of energetic molecules (their energetic transition downward) feeds into the recomposition of other parts of the ecology (the upward transition of energy)" (523).

6 In the brief mentions, the language flouts our contemporary racial categories: "There were smatterings of [the indigenous population], to be sure, pockets of residents on the outskirt of what is now the heart of B-Mor, these descendants of nineteenth-century African slaves and twentieth-century laborers from Central America and even bands of twenty-first-century urban-nostalgics" (Lee 2014a: 19).

7 Heather J. Hicks (2016: 7) writes about postapocalyptic novels characterized by "globalized ruin" that "portray catastrophe of at least a national level and, by nature of our globalized political economy, assume dramatic effects elsewhere."

8 Beyond erasing Indigenous peoples, this slippage of "native" also shows that race is structural position rather than essential embodiment. There is also a Rivera-Deng family in B-Mor, suggesting mixed Latinx or Filipinx ancestry.

9 As children, Freddie Gray and his sisters were found to have damaging levels of lead in their blood as a result of the crumbling paint on the walls of their apartment (Marbella 2015). In the altercation with police leading up to his death, Gray was apprehended by police in the nearby Gilmor Homes public housing complex. Gilmor Homes, which gets its name from Harry Gilmor, a Confederate cavalry officer who would later serve as Baltimore City Police Commissioner, was in such poor condition it was torn down in 2020 (Campbell 2020).

10 Katherine McKittrick (2011: 951) refers to this obliteration as urbicide, or "deliberate death of the city and willful place annihilation" that erases a "black sense of place" (947).

11 Another referent could be Seneca Falls, the site of the 1848 women's rights convention. In the novel, gender norms have not progressed; leadership and head-of-household positions are filled by men.

12 There are researchers pushing back against this postracial imperative— for example, Northwestern University's African American Cardiovascular Pharmacogenomics Consortium (ACCOuNT) focuses on biomarkers found in African American populations to predict responses to cardiovascular drugs.

13 This is thematized incisively in Ken Liu's Hugo- and Nebula-nominated novella *The Man Who Ended History* (2011).

14 In Larissa Lai's *Salt Fish Girl* (2002), a novel that came out a year after the HGP's draft publication, Asian American women are spliced with carp genes to disqualify them as human, showing "the discourse that equates essential humanity with genetic purity is a much more powerful dividing practice than any 0.03-percent snippet of genetic code" (Huang 2016: 132).

15   See Byrnes (2020: 16) on the contradictions of the human/animal divide revealed by swine flu: "while epidemiologists and public health officials continue to publicize the 'pathological state of hybridity' that can result from some kinds of human-animal intimacy, genetic engineers are now actively trying to erase the species boundaries between humans and pigs" (16). *Severance* also features a viral pandemic originating from China called "Shen Fever."

16   Sandra Soo-Jin Lee (2020: S54) writes that "public investment in genomic sciences has contributed to a growing ecosystem of commercial products and services that are no longer considered discretionary or fringe but are instead marketed as key to the maintenance of health for the rational biocitizen."

17   Reg's clan's name is suggestive of the Xinjiang autonomous region in China where many ethnic minorities, including the Uyghur people, reside, as well as of the Xizang autonomous region (Tibet).

References

Banner, Olivia. 2017. *Communicative Biocapitalism: The Voice of the Patient in Digital Health and the Health Humanities*. Ann Arbor: Univ. of Michigan Press.

Barrish, Philip. 2018. "Speculative Fiction and the Political Economy of Healthcare: Chang-rae Lee's *On Such a Full Sea*." *Journal of Medical Humanities* 40, no. 3: 297–313.

Berlant, Lauren. 2011. *Cruel Optimism*. Durham, NC: Duke Univ. Press.

Bonham, Vence L., Shawneequa L. Callier, and Charmaine D. Royal. 2016. "Will Precision Medicine Move Us beyond Race?" *New England Journal of Medicine* 374, no. 21: 2003–5.

Bonilla-Silva, Eduardo. 2017. *Racism without Racists: Color-Blind Racism and the Persistence of Racial Inequality in America*. Fifth edition. New York: Rowman & Littlefield.

Bridges, Khiara M. 2017. "Lessons from Racial Medicine: The Group, the Individual, and the Equal Protection Clause." In *Subprime Health: Debt and Race in U.S. Medicine*, edited by Nadine Ehlers and Leslie R. Hinkson, 155–82. Minneapolis: Univ. of Minnesota Press.

Brink, Susan. 2016. "What's the Prognosis for $3 Billion Zuckerberg Health Plan?" *NPR Goats and Soda*, September 23. www.npr.org/sections/goats andsoda/2016/09/23/495184078/whats-the-prognosis-for-mark-zuckerbergs -3-billion-health-plan.

Brown, Adrienne. 2017. *The Black Skyscraper: Architecture and the Perception of Race*. Baltimore: Johns Hopkins Univ. Press.

Byrnes, Corey. 2020. "Transpacific Maladies." *Social Text* 38, no. 3: 1–26.

Campbell, Colin. 2020. "Five Years after Freddie Gray's Death, Partial Demolition of Gilmor Homes Underway in Baltimore." *Baltimore Sun*, April 17. www.baltimoresun.com/maryland/baltimore-city/bs-md-ci-gilmor-homes -demolition-20200417-l3yuxg4xurattnibpvzna3t46m-story.html.

Carruth, Allison. 2018. "Wily Ecologies: Comic Futures for American Environmentalism." *American Literary History* 30, no. 1: 108–33.

Chen, Mel Y. 2012. *Animacies: Biopolitics, Racial Mattering, and Queer Affect.* Durham, NC: Duke Univ. Press.

Cheng, Anne Anlin. 2020. "Digesting Asian America." In *The Cambridge Companion to Literature and Food*, edited by J. Michelle Coghlan, 215–27. New York: Cambridge Univ. Press.

Chiang, Ted. 2019. "It's 2059, and the Rich Kids Are Still Winning." *New York Times*, May 27. www.nytimes.com/2019/05/27/opinion/ted-chiang-future-genetic-engineering.html.

Chuh, Kandice. 2013. "On (Not) Mentoring." *Social Text Periscope*, January 13. https://socialtextjournal.org/periscope_article/on-not-mentoring/.

Cutler, John Alba. 2015. *Ends of Assimilation: The Formation of Chicano Literature.* New York: Oxford Univ. Press.

Ehlers, Nadine and Leslie R. Hinkson. 2017. "Introduction: Race-Based Medicine and the Specter of Debt." In *Subprime Health: Debt and Race in U.S. Medicine*, edited by Nadine Ehlers and Leslie R. Hinkson, vii–xxxi. Minneapolis: Univ. of Minnesota Press.

Fan, Christopher T. 2017. "Animacy at the End of History in Chang-rae Lee's *On Such a Full Sea*." *American Quarterly* 69, no. 3: 675–96.

Flagg, Anna, Damini Sharma, Larry Fenn, and Mike Stobbe. 2020. "COVID-19's Toll on People of Color Is Worse Than We Knew." *The Marshall Project*, August 21. www.themarshallproject.org/2020/08/21/covid-19-s-toll-on-people-of-color-is-worse-than-we-knew.

Gilmore, Ruth Wilson. 2007. *Golden Gulag: Prisons, Surplus, Crisis, and Opposition in Globalizing California.* Berkeley: Univ. of California Press.

Gupta, Akhil. 2018. "The Future in Ruins: Thoughts on the Temporality of Infrastructure." In *The Promise of Infrastructure*, edited by Nikhil Anand, Akhil Gupta, and Hannah Appel, 62–79. Durham, NC: Duke Univ. Press.

Heise, Ursula. 2015. "What's the Matter with Dystopia?" *Public Books*, February 1. www.publicbooks.org/whats-the-matter-with-dystopia/.

Heng, Rachel. 2018. *Suicide Club: A Novel about Living.* New York: Henry Holt.

Heron, Gil-Scott. 1970. "Comment #1." Track 4 on *Small Talk at 125th and Lenox*. Dutchman Records.

Hicks, Heather J. 2016. *The Post-Apocalyptic Novel in the Twenty-First Century: Modernity beyond Salvage.* New York: Palgrave Macmillan.

Hong, Grace Kyungwon. 2018. "Speculative Surplus: Asian American Racialization and the Neoliberal Shift." *Social Text* 36, no. 2: 107–22.

Houser, Heather. 2014. *Ecosickness in Contemporary U.S. Fiction: Environment and Affect.* New York: Columbia Univ. Press.

Huang, Michelle N. 2016. "Creative Evolution: Narrative Symbiogenesis in Larissa Lai's *Salt Fish Girl*." *Amerasia* 42, no. 2: 118–38.

Hurley, Jessica. 2020. *Infrastructures of Apocalypse: American Literature and the Nuclear Complex.* Minneapolis: Univ. of Minnesota Press.

Hurley, Jessica, and Dan Sinykin. 2018. "Apocalypse: Introduction." *ASAP/Journal* 3, no. 3: 451–66.

James, Jennifer C. 2011. "Ecomelancholia: Slavery, War, and Black Ecological Imaginings." In *Environmental Criticism for the Twenty-First Century*, edited by Stephanie LeMenager, Teresa Shewry, and Ken Hiltner, 163–78. New York: Routledge.

James, Jennifer C. "Dread." 2020. *American Literature* 92, no. 4: 689–95.

Kim, Claire Jean. 1999. "The Racial Triangulation of Asian Americans." *Politics and Society* 27, no. 1: 105–38.

Lee, Chang-rae. 2014a. *On Such a Full Sea*. New York: Riverhead.

Lee, Chang-rae. 2014b. "The Chorus of 'We': An Interview with Chang-rae Lee." Interview by Cressida Leyshon. *New Yorker: Page-Turner*, January 6. www.newyorker.com/books/page-turner/the-chorus-of-we-an-interview -with-chang-rae-lee.

Lee, Rachel. 2018. "Are Biocultural Creatures Posthistorical Agents?" *Theory and Event* 21, no. 2: 518–28.

Lee, Robert G. 1999. *Orientals: Asian Americans in Popular Culture*. Philadelphia: Temple Univ. Press.

Lee, Sandra Soo-Jin. 2020. "Excavating the Personal Genome: The Good Biocitizen in the Age of Precision Health." *The Hastings Center Report* 50, no. S1: S54–S61.

Ma, Ling. 2018. *Severance*. New York: Farrar, Straus and Giroux.

Marbella, Jean. 2015. "Beginning of Freddie Gray's Life as Sad as Its End, Court Case Shows." *Baltimore Sun*, April 23. www.baltimoresun.com /maryland/baltimore-city/bs-md-freddie-gray-lead-paint-20150423-story .html.

McKittrick, Katherine. 2011. "On Plantations, Prisons, and a Black Sense of Place." *Social and Cultural Geography* 12, no. 8: 947–63.

Min, Susette. 2016. "Biopower, Space, and Race in Asian American Studies." *American Literature* 88, no. 4: 839–54.

Mitchell, Robert. 2016. "Biopolitics and Population Aesthetics." *South Atlantic Quarterly* 115, no. 2: 367–98.

Mjoseth, Jeannine. 2016. "Perspective: Precision Medicine May Move Us beyond the Use of Race in Prescribing Drugs." *National Institute of Health, National Human Genome Research Institute*, June 9. www.genome.gov /news/news-release/Perspective-Precision-medicine-may-move-us-beyond -the-use-of-race-in-prescribing-drugs.

Nadkarni, Asha. 2019. "Eugenics, Reproduction, and Asian American Literature." In *Oxford Encyclopedia of Asian American and Pacific Islander Literature and Culture*, edited by Josephine Lee, 1–19. New York: Oxford Univ. Press.

National Research Council. 2011. *Toward Precision Medicine: Building a Knowledge Network for Biomedical Research and a New Taxonomy of Disease*. Washington, DC: National Academies Press.

Nelson, Alondra. 2016. *The Social Life of DNA: Race, Reparations, and Reconciliation After the Genome*. Boston: Beacon Press.

Obama, Barack. 2016. "Medicine's Next Step." *Boston Globe*, July 6. www .bostonglobe.com/opinion/2016/07/06/medicine-next-step/tPdgf4XfOH vUckHpTTbuvN/story.html.

Osnos, Evan. 2017. "Survival of the Richest: Why Some of America's Wealthiest People Are Prepping for Disaster." *New Yorker*, January 30: 36–45.

Palumbo-Liu, David. 1999. *Asian/American: Historical Crossings of a Racial Frontier*. Palo Alto, CA: Stanford Univ. Press.

Park, Robert E. and Ernest W. Burgess. (1921) 1969. *Introduction to the Science of Sociology*. Third Edition. Chicago: Univ. of Chicago Press.

Park, Lisa Sun-Hee. 2015. "Assimilation." In *Keywords for Asian American Studies*, edited by Cathy Schlund-Vials, Linda Trinh Võ, and K. Scott Wong, 14–17. New York: New York Univ. Press.

Puar, Jasbir K. 2009. "Prognosis Time: Towards a Geopolitics of Affect, Debility and Capacity." *Women and Performance* 19, no. 2: 161–72.

Rana, Junaid. 2015. "Race." In *Keywords for Asian American Studies*, edited by Cathy Schlund-Vials, Linda Trinh Võ, and K. Scott Wong, 202–7. New York: New York Univ. Press.

Robbins, Bruce. 2007. "The Smell of Infrastructure: Notes toward an Archive." *boundary 2* 34, no. 1: 25–33.

Rubenstein, Michael, Bruce Robbins, and Sophia Beal. 2015. "Infrastructuralism: An Introduction." *Modern Fiction Studies* 61, no. 4: 575–86.

Saldívar, Ramón. 2013. "The Second Elevation of the Novel: Race, Form, and the Postrace Aesthetic in Contemporary Narrative." *Narrative* 21, no. 1: 1–18.

Sharpe, Christina. 2016. *In the Wake: On Blackness and Being*. Durham, NC: Duke Univ. Press.

Wald, Priscilla. 2008. *Contagious: Cultures, Carriers, and the Outbreak Narrative*. Durham, NC: Duke Univ. Press.

Waldby, Catherine and Robert Mitchell. 2006. *Tissue Economies: Blood, Organs, and Cell Lines in Late Capitalism*. Durham, NC: Duke Univ. Press.

Washington, Harriet. 2020. "How Environmental Racism Fuels Pandemics." *Nature* 581: 241.

Wasserman, Sarah. 2020. "Ralph Ellison, Chester Himes, and the Persistence of Urban Forms." *PMLA* 135, no. 3: 530–45.

Whyte, Kyle Powys. 2017. "Our Ancestors' Dystopia Now: Indigenous Conservation and the Anthropocene." In *Routledge Companion to the Environmental Humanities*, edited by Ursula K. Heise, Jon Christensen, Michelle Niemann, 206–15. New York: Routledge.

Book Reviews

Telegraphies: Indigeneity, Identity, and Nation in America's Nineteenth-Century Virtual Realm. By Kay Yandell. New York: Oxford Univ. Press. 2019. x, 209 pp. Cloth, $78.00; e-book available.

Modernizing Solitude: The Networked Individual in Nineteenth-Century American Literature. By Yoshiaki Furui. Tuscaloosa: Univ. of Alabama Press. 2019. x, 239 pp. Cloth, $54.95; e-book, $54.95.

Gears and God: Technocratic Fiction, Faith, and Empire in Mark Twain's America. By Nathaniel Williams. Tuscaloosa: Univ. of Alabama Press. 2018. xii, 206 pp. Cloth, $44.95; e-book, $44.95.

Adapted from Michael Lewis's book *Moneyball: The Art of Winning an Unfair Game* (2003), the 2011 film *Moneyball,* directed by Bennett Miller, traces the powerful influence of sabermetrics, the empirical analysis of statistics, on baseball by following the fortunes of the Oakland Athletics during a period in which general manager Billy Beane tests the theory held by economics graduate Peter Brand that on-base percentage is a more reliable predictor of player success than batting average or scout evaluations. Brand's theory, adopted from Bill James, relied a good deal on statistical evidence managed by computer programming, and Beane utilized these metrics to determine which players to sign and where to play them. *Moneyball* is a sports story that references technology, innovation, and religious faith as key American values. The volumes under review similarly trace the development of mobile technologies, telegraphy and other communication systems, and the postal system and characterize their influence, evaluating the impact of these technologies on nineteenth-century writers and texts. Yandell, Furui, and Williams discuss how these texts and technologies are shaped by American politics and spiritual values, revealing underpinnings of modernist and postmodern ideas of self and society.

American Literature, Volume 93, Number 3, September 2021
© 2021 by Duke University Press

Telegraphies discusses "telegraph literature," a term Yandell applies to "the fiction, poetry, social critique, and autobiography" produced by writers from "vastly different social locations" in mid- and late-nineteenth-century America (3). She analyzes diverse acknowledgements of "telegraph speech" in canonical texts, including Emily Dickinson's envelope and other poems, Nathaniel Hawthorne's *The House of the Seven Gables* (1851), and poetry by Walt Whitman. Noting that telegraphy carries consciousness across wires, Yandell also looks at Native American telecommunication practices. In a fascinating chapter titled "Moccasin Telegraph," she lists an abundance of visual communication methods (pictographs, petroglyphs, sign language, quipus, ceremonial dance, codices, smoke signals, and alphabetic writing) as a context for discussion of Frank Linderman's interviews in the 1920s with two tribal members, Plenty Coups and Pretty Shield, who used sign language to tell Linderman stories about traditional Crow communication methods. Another chapter examines gender, race, and class issues raised in fictional and autobiographical narratives about and by telegraph clerks. Yandell's conclusion describes the way that *The House of the Seven Gables* "addresses visions of history as an evolutionary spiral through competing social theories of the telegraph" (158), which is imaged as both an exciting advancement and a look back at a past burdened with moral guilt.

Communication and transportation technologies both connect individuals and emphasize their alienation from one another, a situation paralleling contemporary experiences of the Web and social media: one may be connected over Facebook or Twitter or Instagram with hundreds or thousands of individuals who live all around the world, yet one can still feel lonely. Furui's *Modernizing Solitude* addresses concepts of networking and solitude that arise subsequent to the implementation and expansion of innovations that "reshaped concepts of spatiality and temporality": "the Postal Acts of the 1840s and 1850s, the invention of the telegraph in the 1840s, and beginning in the 1830s, the development of the railroad, the Fourdrinier printing process, cheap paper making, and other technological innovations" (1). Furui develops an argument about the modern concept of solitude by tracing its development in works by Henry David Thoreau, Harriet Jacobs, Herman Melville, Emily Dickinson, and Henry James. Covering "the period from 1831, the year that William Lloyd Garrison founded his antislavery newspaper, the *Liberator*, to 1898, when James published 'In the Cage'" (19), Furui explains the ways each writer found the virtues and values of solitude to be critical in identity formation. *Walden* (1854) reveals that Thoreau, "a pseudo-hermit" (33), craved solitude while dreading isolation; he remained connected to a wider world as he read newspapers and made daily visits to family and friends while living at Walden Pond, where he could daily hear the newly built rail system connecting towns across Massachusetts. Jacobs endures "networked solitude" (47) in *Incidents in the Life of a Slave Girl* (1861), describing a captivity during which she received and relayed messages to send miscues to her pursuers. Amherst residents, including the Dickinson family, were well aware of changes wrought by rail and telegraph. Applying Benedict Anderson's term "imagined communities," Furui traces Dickinson's engagement with these technologies and with newspapers as revealed by references in her poetry. Furui contrasts Melville's story world with such illustrations of networked solitude, finding the lawyer's office in "Bartleby, the Scrivener" (1853) to represent instead "a culture of dead letters"; Bartleby's unhealthy, motionless solitude serves as a form of resistance to "the world set in flux by the communications revolution" (69). Telegraphic fictions such as James's novella *In the Cage* (1898) reveal the technology's inability to increase class mobility, as operators find it difficult to attain the same status as their customers. A similar limitation applies to contemporary ambivalence about the powers of artificial intelligence, as Furui's epilogue considers in outlining the impossibility of true intimacy in the human-cyborg romance represented in *Her*, a 2013 film directed by Spike Jonze.

In *Gears and God*, Williams argues that "technocratic exploration tales" include discussions of science and technology as a way to manage questions of faith. Technocratic literature "allowed fantasies of technological empowerment, but without a consistent endorsement of empire" (17) at a time of skepticism concerning Darwinian evolutionary theory. Williams finds common themes in a subgenre of hundreds of dime adventure novels incorporating technologies as solutions to challenges, for these works and many science fictions consider what late-nineteenth-century audiences regarded as "core American values: pro-technology, pro-expansion, and pro-Christianity" (4). *Gears and Gods* points to connections between technology and expansionism in early American science fiction, in "Edisonades" (stories featuring inventors), and in contemporary media. One chapter looks at the topic of biblical literalism and other patterns in Frank Reade novels as background to a study of correspondence between Samuel Clemens and Orion Clemens that discusses elements of a proposed novel by the less successful older brother, who aimed to improve on Jules Verne's fictions. Williams points out that Samuel Clemens (Mark Twain) used personal narratives of frontier life and adventure as source material for *Huck Finn and Tom Sawyer among the Indians* (1889), a novel that "used frontier exploratory tropes and assessed the role of religion and technology in facilitating such expansion" (111) and that served as a precursor for *A Connecticut Yankee in King Arthur's Court* (1889) and *Tom Sawyer Abroad* (1894).

Yandell invokes the term "virtual realm" to point out the parallels between nineteenth-century technologies connecting people and contemporary communication forms enabling connection or highlighting disconnection, and it is a common theme among these volumes that cultural reactions to nineteenth-century technological innovations resemble responses to the Internet in our own day. Perhaps it is inevitable that contemporary experiences with the Web and other technologies inspired these three authors to look back at how American literature came to terms with the communication and mobile technologies of the nineteenth century. Accessibly written and offering analyses of texts and contexts, these books should be useful for students and scholars interested in the nineteenth-century communications and transportation

revolutions as an analogue to a more recent paradigm shift to digital communications shadowed by pandemic and lockdown in our own time.

Carol Colatrella is professor of literature; codirector of the Center for the Study
of Women, Science, and Technology; and associate dean for graduate studies and
faculty development at Georgia Institute of Technology. She has published *Evolution,
Sacrifice, and Narrative: Balzac, Zola, and Faulkner* (1990); *Literature and Moral
Reform: Melville and the Discipline of Reading* (2002); and *Toys and Tools in Pink:
Cultural Narratives of Gender, Science, and Technology* (2011), and she has edited
Technology and Humanity (2012) and coedited (with Joseph Alkana) *Cohesion and
Dissent in America* (1994), essays written to honor Sacvan Bercovitch.

DOI 10.1215/00029831-9361307

White Writers, Race Matters: Fictions of Racial Liberalism from Stowe to Stockett.
By Gregory S. Jay. New York: Oxford Univ. Press. 2017. xi, 370 pp. Cloth, $73.00;
e-book available.

Black Prometheus: Race and Radicalism in the Age of Atlantic Slavery. By Jared
Hickman. New York: Oxford Univ. Press. 2016. x, 528 pp. Cloth, $81.00; paper,
$39.95; e-book available.

The two books under review explore how the racial formations of background
epistemologies and ideologies shape and have shaped cultural formations in
the Anglosphere. *White Writers, Race Matters* largely confines its analytical
remit to twentieth-century "white-authored protest fiction about racism in
America," attempting to come to grips with the endurance of liberal race novels, and racism liberalism, in the United States (3). *Black Prometheus*, meanwhile, is a book about nearly everything—or, at least, everything after 1492.
Hickman's ranging, audacious volume approaches Romantic-era (capaciously
defined) conflicts over slavery, abolition, colonialism, and racialization from a
postsecular perspective, urging us to see these antagonisms as instancing a
cosmological struggle among competing conceptions of divinity, and, indeed,
as a struggle among gods themselves.

   White Writers, Race Matters takes up white-authored liberal race fiction as a
literary tradition, and a "self-conscious tradition" at that (6). For Jay, this literary tradition addresses itself to a largely white audience, and, from Harriet
Beecher Stowe on, it aims to "educat[e] the hearts, and propel[] the actions,
of those who think they are white" (6). Jay's six chapters mix the still popular
and the now obscure, with Fannie Hurst's *Imitation of Life* (1933), Harper
Lee's *To Kill a Mockingbird* (1960), and Kathryn Stockett's *The Help* (2009) sit-
ting alongside Laura Z. Hobson's *Gentleman's Agreement* (1947) and Lillian
Smith's *Killers of the Dream* (1949). Moving across these texts and their many
adaptations into film, Jay continually lands a simple but effective point: the
cultural genres of white racial liberalism are popular, and profitable.

In Jay's reading, the attractions of racial liberalism for white audiences derive precisely from racial liberalism's ideological limits. For Jay, racial liberalism names an ideological disposition in which racial inequality is primarily addressable as a failure for whites to feel right about racial others; liberal race fiction attempts to cultivate readers' capacities for sympathetic relations to racial others. The political promise of racial liberalism is that feeling right will lead to doing right, that a change of heart will lead to transformations of structured inequalities. But, as Jay concedes, and as the body of critical scholarship on race, liberalism, and sentimentality has long attested, this promise rarely cashes out: feeling right winds up feeling as if it were itself a form of political action. What distinguishes *White Writers, Race Matters* from this critical tradition (as simple signposts, think of Lauren Berlant's *The Female Complaint* [2008] and Jodi Melamed's *Represent and Destroy* [2011]) is Jay's ambivalence regarding racial liberalism's self-canceling relation to political action. For Jay, the works he explores are always more complicated, and more complicating, than critical dismissals of them would allow. Citing Rita Felski on "the limits of critique," Jay insists that he does not want to "write an apology" for, in this instance, *The Help* but rather to recover "other fruits of inquiry made possible by a more generous and less judgmental attitude toward the book and its author" (290). Yet, what Jay poses as a minimization of judgment is rather an evaluative reorientation; his posture of ambivalence entails (and indeed requires) a great deal of recuperation. All of the texts Jay explores are politically compromised or toxic in some ways, as he acknowledges—only then to counter the failures of or problems with any given text's execution against the rectitude of the text's intention. White-authored liberal race fiction operates by reconciling white audiences to a racial positioning it can critique but cannot abolish; feeling right—which might be to feel guilty, critical, implicated, sympathetic—becomes an actionless but virtuous end in itself. In many ways, *White Writers, Race Matters* is a text convincing itself to feel okay about white liberalism's popular, profitable, frequently racist, and never transformative feelings work.

Where Jay's book takes up white racial liberalism, Hickman's *Black Prometheus* takes on an epistemological, ontological, and theological disposition that might once have been called secularism. For Hickman, secularist accounts of modernity (in particular, those of Hans Blumenberg) misperceive the conditions of their possibility and in the process reproduce white and European supremacist structures of knowing. In Hickman's account, the emergence of modernity is tethered to 1492, an event that positioned the "shifting relations of former cosmic others within the now single, dynamic cosmos of the planet," which Hickman also refers to as the "immanent frame of the globe" (42). Secularist accounts of modernity, Hickman argues, adduce the immanent frame of the globe as a precondition for secularism; on this read, the unbounded contact of diverse cosmologies relativized each cosmology, even as globalization effected a detranscendentalization of the cosmos and a terrestrialization of the gods. Against this narrative, Hickman argues that "the immanent frame of globality conduces not so much to the death of God as to a new kind of life for gods, one in which gods and their human bearers become conflated in ways that both humanize divinities and divinize humans" (45). The gods are brought down to earth, and the violent conflicts attending European expansion are in fact battles between their gods and their human devotees. From this perspective, modernity "is structured not by the agon between God and Man," as a secularist account would hold, taking modernity as the emancipation of humanity from theological and religious irrationality, transcendence, and so forth; rather, modernity is structured "by the agon between apotheosized Euro Christian and anathematized non-Euro-Christian heather" (61). Secularist thought takes modernity to be a process wherein European humanity discards God; whereas, for Hickman, modernity amounts to the "Euro-Christian's *becoming* God vis-à-vis non-Euro Christian heathens" (61). Hickman organizes his inquiry into the philosophical, cultural, and cosmic life around two different figurations of Prometheus. First, there is the white, Euro-Christian, pseudo-secular Prometheus, the one who stole from the gods to liberate humanity. Against this, the archive that Hickman populates through the trope of Black Prometheus attends to the persistence of the divine in the immanent materiality of human life, a persistence of other cosmologies set to work by the enslaved and the colonized in their war against Christian human-gods.

I do not mean this as a critique: Reading *Black Prometheus* is like tripping on acid. It is an immersive experience, one that defamiliarizes what and how we know the world. Indeed, Hickman considers this book to be a "meta-cosmography," inasmuch as it aims to discern "the mapping and shaping of what has become our specific terrain of the universal, the planet we all inhabit, with its ever-densifying history" (18). Put differently, *Black Prometheus* fully inhabits the cosmos of warring gods and god-men that it describes. This commitment to the bit yields some incredibly dazzling, inventive readings. Hickman's third chapter, for instance, "Africa versus the Absolute: Idealism and Its Others," reads Frederick Douglass's *My Bondage and My Freedom* (1855) with and against Hegelian and Marxist conceptualizations of materialism, the Absolute, and human freedom. Hegelian readings of *My Bondage* abound; Douglass's fight with Covey basically cries out to be read alongside Georg Wilhelm Friedrich Hegel's *Phenomenology* (1807). In one of the many brilliant turns of the chapter, Hickman argues that "the place to look for a new history of freedom from the perspective of the black Atlantic is not in the lord-bondsman dialectic but rather in the metadialectic space between Hegel's Absolute and Hegel's "Africa" (128). If Hegel's "Africa" is cut off from his Absolute, and mired in particularity, nonideality, and materiality, Hickman argues, *My Bondage* radically potentiates that (otherwise abject) positioning in the service of generating a thought of freedom that is radically released from any dependence upon an absolute God or Master (150). His other chapters— on Romantic Prometheanism, Black astronomy, Mary Shelley's *Frankenstein* (1823) and Ralph Waldo Emerson, "Byronic Abolitionism," and more—are equally brilliant, probing, and inventive.

Black Prometheus succeeds as a fully globalized (indeed, cosmic) account of diverse Atlantic and American literary formation—one for which globality isn't presenced by selecting texts from here and there but is rather an episte-mological and theoretical attunement, a mode of emplotting and thinking. Hickman allows us to see how race *is* a theology, and one with its own distinct cosmography. *Black Prometheus* is a doorstopper, and some readers may be frustrated by its highly theoretical idiom, but its length and difficulty are more than well worth it.

Chris Taylor is associate professor of English at the University of Chicago and the author of *Empire of Neglect: The West Indies in the Wake of British Liberalism* (2018).

DOI 10.1215/00029831-9361321

Cavaliers and Economists: Global Capitalism and the Development of Southern Literature, 1820–1860. **By Katharine A. Burnett. Baton Rouge: Louisiana State Univ. Press. 2019. xi, 266 pp. Cloth, $49.95.**

Autonomy: The Social Ontology of Art under Capitalism. **By Nicholas Brown. Durham, NC: Duke Univ. Press. 2019. x, 219 pp. Cloth, $99.95; paper, $25.95; e-book available.**

Katharine A. Burnett opens *Cavaliers and Economists* with the premise that southern literature developed "in tandem with economic modernization—not in spite of it" (4). She goes on to suggest that, even as early as the 1830s, southern literary efforts "developed in resistance to shifts in an emerging cap-italist system and used those shifts to rationalize slavery and southern society as it existed" (4). Burnett's work is clearly influenced by recent attempts to approach southern studies through a global lens, yet her insistence that earlier southern literature embraces, rather than resists, global capitalism becomes especially interesting when she suggests that nineteenth-century southern writers used "the language of economics . . . to reframe mainstream representations of the region through popular literary forms and genres that proliferated in the early-nineteenth century transatlantic literary culture" (9). Hinting at a similarity between capitalist systems' use of the marketplace to generate and sustain economic values and the way fiction uses literary form and genre to define cultural values, Burnett's argument takes us in a very interesting direction right away.

Approaching southern literature as part of a broader artistic movement, the book has five main chapters, each of which interprets southern writing "against the work of British authors who shape the core of the nineteenth-century literary canon" (16). The book addresses a range of literary genres, however Burnett's reach extends well beyond the aesthetics of form, offering insights into the way these texts use the language of economics (most often the language of capitalism) in order to "create representations of the region in which liberal capitalism could coexist with the socially regressive sites of the antebellum plantation" (4). While she keeps a close eye on the individual texts, her chief aim— the argument that this writing developed in response to economic and cultural shifts throughout 1800s—is never far away.

Pursuing a global perspective, by necessity, makes the argument expansive, but Burnett weaves the threads tightly. A particularly nice argument emerges in the fourth chapter, which addresses southern writers' appropriation of the social problem model, in which characters are powerless to change the socioeconomic forces that shape their lives. She argues, for example, that Donald Montrose, the protagonist of Maria Jane McIntosh's proslavery *The Lofty and the Lowly* (1853), demonstrates a financial recklessness that "mirrors the pattern of the average plantation owner described by [George] Tucker in *The Valley of the Shenandoah*, in which southern aristocrats squander money with the promise of future returns on crops" (164). But Tucker's earlier book, she contends, tries to strike a more delicate balance between "overt criticism of slavery and reluctance to dismantle the system [and] adopts the form of the eighteenth-century sentimental novel of seduction and applies it to economic commentary" (138). In other words, these narratives imagine southern aristocrats who acknowledge (and perhaps even suffer) the ills of slavery but are powerless to change the social and economic systems in which they are entangled. Observing that southern writers could repurpose this English model to "project a version of reform in their fiction," Burnett invokes certain global literary relationships, but the argument gets especially appealing when she points out that along with their domestic designs, these novels reveal a more profound purpose, which is "writing the slaveholding south into a narrative of global economic progress" (136). Bringing the point into full focus, she claims that this literary exchange is analogous to more literal economies and that ideologies shaped by those economies (economic and literary) led to nineteenth-century southern writers adopting "the language of economics [which] allowed them to reframe mainstream representations of the region through popular literary form and genres that proliferated in the early-nineteenth-century transatlantic literary culture" (90). In other words, those shared forms reveal that southern writers were actively working to normalize their "peculiar institution" by writing it into the fabric of global economies, both literary and financial.

Nicholas Brown's *Autonomy: The Social Ontology of Art under Capitalism* is similarly concerned with the ways art is entangled in questions of value, but where Burnett is focused on its social capital, Brown's attention in on the question of how the art can exist in a capitalist system where all value is defined by the exchange of commodities. The scope of *Autonomy* is so wide that almost anything said about it will be an oversimplification. Brown's argument is wide-ranging but comprehensive. He delves into, among other things, aesthetics, Marxism, philosophy, art criticism, and music theory in the service of interrogating the question of how art generates aesthetic value while sometimes having commodity value. The problem, as Brown sees it, is that if "a work of art is not only a commodity—if a moment of autonomy with regard to the commodity form is analytically available . . . then it makes entirely good sense to approach it with interpretive tools," but "if a work of art is only a commodity, interpretive tools suddenly make no sense at all" (8). In some ways, Brown's argument boils down to the question of what the difference is between a work of art that is performing its value and a work of art *aware* that it is performing its value. Brown's definition of autonomy is, by necessity, a bit foggy, but ultimately a work of art's interaction with social externalities is key, and art's inescapable *thingness* will always complicate that.

Brown's argument takes flight in chapter 3, specifically in his discussion of Bertolt Brecht's *The Three Penny Opera* (1928), which he believes illustrates his point about commodity and splitting the difference between means and end in didactic theater. Personal affinity as well as a certain academic wonder compels me to point out that it's here Brown makes a remarkably cogent point about modern pop music while calling back to *The Three Penny Opera*, the White Stripes, and Claude Debussy. In fact, one of the things that sets this monograph apart is its commitment to looking at work from a remarkably wide range of social and aesthetic strata in order to illustrate the fullness of his approach.

Ultimately, both these studies have provocative things to say about the ways we think about art and its relationship to value. Where Burnett's argument points toward an understanding of antebellum southern literature as something that expresses its autonomy when it "exposes the interconnected nature of literary form and economic change and lays bare the centrality of slavery and regional identity in global capitalism" (206), Brown's argument winds a different course, ultimately concluding that contemporary global capitalism has become the very fabric of our understating of value in all its forms, which means even as art confronts it, capitalism transforms art into "a consumable sign of opposition" (182). Ultimately, the conversation Burnett and Brown are having is about understanding the more complex ways art and economics interact and whether the question of aesthetic value can ever be understood outside the context of economic value.

Christopher Bundrick is an associate professor of English at the University of South Carolina Lancaster. His research focuses on southern regionalism, and he is the author of essays about Thomas Nelson Page, Mary Murfree, Charles Chesnutt, and Elliott White Springs.

DOI 10.1215/00029831-9361335

Tombstone, Deadwood, and Dodge City: Re-creating the Frontier West. By Kevin Britz and Roger L. Nichols. Norman: Univ. of Oklahoma Press. 2018. xiii, 266 pp. Cloth, $32.95; e-book available.

Fictions of Western American Domesticity: Indian, Mexican, and Anglo Women in Print Culture, 1850–1950. By Amanda J. Zink. Albuquerque: Univ. of New Mexico Press. 2018. xiii, 339 pp. Cloth, $75.00; e-book, $75.00.

Failed Frontiersmen: White Men and Myth in the Post-Sixties American Historical Romance. By James J. Donahue. Charlottesville: Univ. of Virginia Press. 2015. viii, 222 pp. Cloth, $59.50; paper, $27.50; e-book, $27.50.

Frederick Jackson Turner famously declared the United States frontier closed in 1893, but of course he was wrong: you can't close an idea. As many historians, literary scholars, and cultural critics since Turner have suggested, the frontier is not a place in time but a grand mythology, a story of national formation in which violent conflict gives way to white patriarchal authority and the domestication of "wild" spaces and people. Rehearsed in every form of art and popular culture, frontier mythology is also a broken promise, since the moral order, homogeneous social accord, and economic prosperity it ascribes to white settlement never arrived. As such, it prompts Americans' continuous engagement; we return to it again and again, either to reinvest in its premises or, more productively, to reevaluate and challenge them. Though they differ in the depths of their investigations, the books under review join the ongoing project of tracking that engagement in the modern era.

Kevin Britz and Roger L. Nichols examine how the towns of Tombstone, Arizona, Deadwood, South Dakota, and Dodge City, Kansas embraced and marketed their histories of frontier violence in the twentieth century. Working with media reports, public records, local histories, and cultural materials, the authors find that civic leaders eager to rebrand their towns initially struggled with the implications of celebrating gunfights and lawlessness while trying to attract new businesses and families. They also faced the problem of representing an authentic frontier past that would at the same time approximate the Wild West of fiction and film that visitors would expect. Ultimately, the three towns went all in on quasi-historical representations of "Shootin'—Lynchin'—Hangin'," to quote a 1947 advertisement for Tombstone's Helldorado Days festival (131). Britz and Nichols document the hucksterism that turned the towns into tourist attractions, with monuments, museums, and "gun and fun" events emphasizing a partially fabricated past (145). But they leave unexplored the broader political, racial, and gendered significance of this choice, not addressing its reinforcement of patriarchal structures of power, its perpetuation of Native American stereotypes, and its effective erasure of African American, Mexican American, Asian American and other populations in the communities. The authors note that the frontier rebranding was motivated in two of the towns by severe economic crisis, but they do not comment on the fact that the current per capita income in all three remains well below the national average—an indication that the myth has twice failed to deliver on its promise of prosperity. How does the commodification of frontier mythology perpetuate the economic inequity of late-stage capitalism? What is the effect of civic endorsement of violent, masculinist, racist mythology on the real people inhabiting these towns? Disappointingly, the authors do not ask.

Amanda J. Zink takes a more critical approach in her focus on the work that frontier mythology assigns to women: cultivating homes in the wild. Zink starts by examining how the white women writers Willa Cather, Edna Ferber, and Elinore Cowan Stone fell in line with popular magazine and advertising practices by delegating domestic labor to the racial other. Their depictions of women of color laboring in white homes cohere with the rhetoric of educators at the Carlisle Indian Industrial School, who used school newspapers to train Indigenous girls in Euro-American domesticity, and with a series of children's novels by Evelyn Hunt Raymond in which white heroines help their Native American, Mexican, and Spanish friends become better housekeepers. Zink juxtaposes such white-authored "colonial domesticity" with "sovereign domesticity," in which Native American women writers appropriate the conventions of domesticity and sentimentality in order to reclaim authority over their own homes. Between these two positions are situated Mexican American women writers María Amparo Ruiz de Burton, Jovita González, Cleofas M. Jaramillo, Fabiola Cabeza de Baca Gilbert, and Nina Otero-Warren, who, Zink argues, negotiate a domestic space between colonized and colonizer: they condemn the racism of white domestic reformers but also sometimes use evidence of their own refined domesticity to claim whiteness. Ranging widely across genre and form, Zink calls attention to little-known novelists, magazine writers, and children's authors from the late nineteenth and early twentieth centuries. She opens up a rich trove of materials that includes cookbooks, housekeeping manuals, Carlisle student writing, and the rhetoric of reformist "better babies" campaigns, her archive itself becoming a frontier space of encounter. Yet despite its efforts to form "textual conversations" between Euro-American, Native American, and Mexican American women, the study is surprisingly unbalanced (23). The most developed close readings are devoted to the three white women authors in the first chapter, the longest in the book. The final chapter's readings of work by the Native American writers Sarah Winnemucca, S. Alice Callahan, Zitkála-Šá, Mourning Dove, and Ella Cara Deloria are altogether too brief. In an intriguing epilogue, Zink introduces fashion as another area in which women of color negotiate gendered colonialism. Moving from the pages of a turn-of-the-twentieth-century magazine edited by Ora V. Eddleman Reed, a writer of Cherokee descent, to Felicia Luna Lemus's 2003 Chicana lesbian coming-of-age novel *Trace Elements of Random Tea Parties*, she exposes fertile new ground for intersectional approaches to the frontier.

The most satisfying of the three studies is James J. Donahue's examination of post-1960s historical romances, largely because it emphasizes the flawed premises and unfulfilled promises of frontier mythology. For Donahue, the mythology betrays us in two ways, first by setting an impossible standard for American masculine identity and then by failing to accommodate the nation's cultural complexity and conflict. He reads historical novels by a group of male writers—E. L. Doctorow, John Barth, Thomas Pynchon, Ishmael Reed, Gerald Vizenor, and Cormac McCarthy—as necessary reevaluations and revisions of the frontier myth in the face of the social and political contexts of the authors' own period, including Vietnam War protests, the civil rights movement, and feminist activism. The desultory, cowardly, sometimes immoral protagonists Donahue examines are antiheroes, men who fail to live up to the idealized image of frontier masculinity posed by figures like Daniel Boone, and in their failure they call attention to the dishonesty of that model. These figures succeed, however, in revealing the serious defects in frontier mythology: its grounding in the authority of histories, first-person narratives, maps, and other documents that are biased or wholly invented; its inherent racism and sexism; its valorization of unspeakable violence; and, not least of all, its tendency toward madness and chaos, rather than order. Donahue's analyses are sharp and his comparisons generative, and while he is not the first to deconstruct the frontier myth in this way (he is indebted to Richard Slotkin, in particular), his reframing of postmodernist novels as historical romances—works that imaginatively grapple with received history and cultural inheritance—is compelling.

Donahue's coda emphasizes the destructive nature of frontier mythology, especially its "dangerously dichotomous separation of peoples," suggesting that the narrative in which the nation has so heavily invested has not just let the nation down but badly broken it (165). Indeed, the mythology continues to inflict real damage. Britz and Nichols describe the rise of social clubs in Tombstone, one of them called "the Vigilantes," dedicated to historical reenactments of shootouts and lynchings. They do not address a former Tombstone reenactor's founding of a civilian militia group, the Minuteman Civil Defense Corps, that patrolled the US-Mexican border in the early 2000s and detained migrants at gunpoint. This present-day vigilantism brings (back) to real life the frontier violence that is only play-acted for tourists. Such instances demonstrate how the unexamined and ritualized myth inevitably converts to truth, in iterations that imperil our existence as a democratic society. Adding to an already rich field of critical inquiry, these books remind us of the imperative to understand one of our most prominent national myths, to recognize its self-perpetuating destructiveness, and to appreciate the ways it has been contested and revised, especially by those it most tyrannizes.

Janet Dean is professor and chair of English and cultural studies at Bryant University, where she writes and teaches about Native American literature, nineteenth-century women writers, and the cultures of political and social protest in the United States. She is the author of *Unconventional Politics: Nineteenth-Century Women Writers and U.S. Indian Policy* (2016), as well as essays, chapters, and reviews in a number of journals and collections. Her current work explores the significance of material culture in Native American literature.

DOI 10.1215/00029831-9361349

Brief Mention

General

John Berryman and Robert Giroux: A Publishing Friendship. By Patrick Samway SJ. Notre Dame, IN: Univ. of Notre Dame Press. 2020. xv, 247 pp. Cloth, $45.00; e-book, $35.99.

This study examines the long, turbulent, personal, and professional relationship between the editor Robert Giroux and the poet John Berryman. Beginning in 1932 (when the two men met at Columbia), the book covers the poet's time at Cambridge and Princeton, the editor's work at Harcourt, Brace and Co. and Farrar, Straus and Cudahy, and both of their scholarship on William Shakespeare. Samway—who worked with Giroux and had access to his letters—concludes with a detailed narrative of Giroux's work on Berryman's final poetry collection (one of dozens he edited) and a summary of Berryman's 1972 death and Giroux's demise in 2008.

Poetry in a Global Age. By Jahan Ramazani. Chicago: Univ. of Chicago Press. 2020. 323 pp. Cloth, $95.00; paper, $30.00; e-book available.

This study examines how modern and contemporary poetry participate in and reflect different forms of "globality," globalism, and "planetary enmeshments." Deploying a variety of methodologies, including cultural geography, ecocriticism, and lyrics studies, the book configures poetry's work in the global age as polyspatial, polytemporal, global, and local. This configuration emerges with the examination of the work of a diverse group of poets—like W. B. Yeats, Wallace Stevens, Lorine Niedecker, Daljit Nagra, and Lorna Dee Cervantes—whose work exemplifies the "transnational dimensions of modern and contemporary poems."

American Literature, Volume 93, Number 3, September 2021
DOI 10.1215/00029831-9361363   © 2021 by Duke University Press

Inter-imperiality: Vying Empires, Gendered Labor, and the Literary Arts of Alliance.
By Laura Doyle. Durham, NC: Duke Univ. Press. 2020. xi, 378 pp. Cloth, $109.95;
paper, $29.95; e-book available.

This study offers a transhistorical, interdisciplinary, intersectional, and
decolonial analysis of the fundamentally relational processes that constitute
imperial powers and individual lives. Polities and persons alike are enmeshed
in shifting entanglements that enable coercion and violence as well as care and
community. Aiming to "honor the struggles and the sustaining practices" that
are elided when this existential interdependence is disavowed, Doyle chronicles a *longue durée* of dialectical state and identity (co)formation that spans the
eleventh to the twentieth centuries. Precapitalist and non-European at its roots,
this archive reorients the study of history, political economy, and critical theory. Literary scholars will also note Doyle's intervention in debates about the
depoliticization of "World Literature," as well as her focus on the formative
power of the aesthetic as it variously furthers and frustrates the status quo.

Love's Shadow. By Paul A. Bové. Cambridge, MA: Harvard Univ. Press. 2021. xi, 431
pp. Cloth, $59.95.

This study seeks to unsettle what it identifies as a dominant academic and
intellectual trend: the tendency to dwell in melancholy and allegoresis. Offering poesies as an alternative, Bové begins with a genealogy of "messianism,
apocalypse and allegory" before arguing that works by Wallace Stevens, Rem-
brandt, and William Shakespeare allow intellectuals to "embrace, defend, and
learn from poetry and criticism." The book ends by considering how the essay
form can be used to eschew acquiescing to dominance in favor of embracing
"the possibility imagination offers."

Hooked: Art and Attachment. By Rita Felski. Chicago: Univ. of Chicago Press. 2020. xiv,
199 pp. Cloth, $95.00; paper, $22.50; e-book available.

Felski's earlier book, *The Limits of Critique* (2015), challenged critical preten-
sions to disinterested interpretation. *Hooked* continues this "postcritical project," exploring the various "attachment[s]" that bond an artwork to its audiences and the wider world. Adapting sociologist Bruno Latour's actor-network
theory, Felski "slow[s] down judgment in order to describe more carefully
what aesthetic experiences are like and how they are made." "Bonds," here, are
"more than restraints," since they "create and make possible" unpredictable
and indeterminate encounters. Endeavoring to "walk around" or "circumvent" stubborn oppositions (between art and society; trained and lay readers;
subjective and objective meaning), *Hooked* devotes chapters to, respectively:
attachment; attunement; identification; and interpretation.

Writing and Righting: Literature in the Age of Human Rights. By Lyndsey Stonebridge.
Oxford: Oxford Univ. Press. 2021. xii, 147 pp. Cloth, $25.00; e-book available.

Addressed to a contemporary global moment in which human rights are
undervalued and under attack, this book explores the connections among literary expression, moral judgment, and political action. Literature must offer
more than a sentimental education, Stonebridge argues, because feelings of
sympathy, pity, and compassion rarely translate to the messy work of enacting
change. Instead, literature should wield its demystifying and creative force in
order to comprehend injustice—to grasp its insidious effects via the insider
testimony of the dispossessed. It is this world-reshaping capacity of art that
Stonebridge locates in a diverse arc of writers that includes Virginia Woolf,
Behrouz Boochani, Suzanne Césaire, Simone Weil, Samuel Beckett, Kamila
Shamsie, Ben Okri, Yousif M. Qasmiyeh, and Hannah Arendt.

Collection

The Colored Conventions Movement: Black Organizing in the Nineteenth Century.
Edited by P. Gabrielle Foreman, Jim Casey, and Sarah Lynn Patterson. Chapel Hill: Univ.
of North Carolina Press. 2021. xxiii, 363 pp. Cloth, $95.00; paper, $29.95; e-book,
$24.99.

This essay collection is the first to examine the Colored Conventions movement, a Black-led network of collective organizing that gathered tens of thousands of participants at dozens of meetings across the United States from
1830 to the turn of the century. Reclaiming this neglected history, the volume
assembles an interdisciplinary range of scholars to map the movement's
myriad meanings for nineteenth-century Black culture and politics. The
resulting study is expansive and rich in detail, especially when paired with
ColoredConventions.org, a collaborative website that houses additional
essays, multimedia exhibits, and datasets. In print and online, this ongoing
project aims in particular to center the oft-ignored but invaluable work of
Black women.

Announcement

Call for Papers—Special Issue of *American Literature*: Senses with/out Subjects

In the past two decades, a perhaps paradoxical development has arisen in literary studies: renewed attention to the human body, in all its perceptual and affective complexities, and the broad rejection of the human body as the basis of epistemological and metaphysical inquiry. Keyed to signification and citation, the "linguistic turn" had the perhaps unintended effect of occluding the body at the core of its analysis of power and knowledge. Today, a range of exciting new scholarship bears out a sustained effort to undo the "sensory deadening" of poststructuralism (cf. Dana Luciano, "How the Earth Feels" [2015] in *Transatlantica*), taking the sensory body as something of the locus classicus of culture, belief, kinship, and power. All the while, posthumanism and its adjacent schools of thought—which describe agency as distributive, ontology as relational, and human and nonhuman matter as entangled or intra-active—have challenged long-held assumptions in literary criticism about the human as the necessary conceptual framework for tracking the movement of texts, ideas, economies, and imaginaries across space and time. At the especially fruitful nexus of critical race theory and posthumanist critique, thinkers such as Mel Y. Chen, Fred Moten, Leanne Betasamosake Simpson, and Sylvia Wynter have unspooled one of the Enlightenment's more insidious legacies: the colonial figure of Man as a rational being capable of claiming sovereignty and superiority over the object world. In Moten's enduring account of the Black radical tradition in *In the Break* (2003), it is through sensory registers—a scream, a caress, a scent—that "objects can and do resist."

Attuned to this recent critical movement from subject to object and back again, this special issue of *American Literature* seeks not to map out another critical "turn" but to build upon and extend the critical possibilities—sparked by the interdisciplinary fields of sensory studies, Black studies, Indigenous studies, disability studies, and feminist, queer, and queer of color theory—that the senses offer for American literary criticism. At once central to the

American Literature, Volume 93, Number 3, September 2021
DOI 10.1215/00029831-9361377 © 2021 by Duke University Press

disciplining of people and populations and entirely irreducible to a single being or body, sensation is a useful heuristic for clearing conceptual space for the human subject, denatured yet still so crucial to the operation of biopower and the asymmetrical effects thereof (across race, gender, class, ability, and species). This issue, then, is an occasion to sketch out the kind of literary histories that come to the fore when we take sensation, in all its diffuse materiality, as an elemental switch point between subject and object worlds. Some questions we ask: What do the senses make of the subject's matter in a more-than-human world? How might a "sensory enlivening" of literary criticism attune us to those writers, thinkers, and artists who have sought to negotiate the human while negating humanism? If sensation operates as a crucial point of relay between subjects and objects, what is the role of the literary (printed, performative, or otherwise) and the aesthetic more broadly in mediating if not amplifying sensation's worldmaking possibilities?

We invite essays that might address any of the following topics:

- Sensory environments (soundscapes, smellscapes, etc.)
- Technologies and techniques of perception
- Aesthetics and cross-sensory modalities (synesthesia, haptics, etc.)
- The regulation of the senses, via settler colonialism, imperialism, racial capitalism, biopower
- Non-Western, non-Protestant, nonableist, Indigenous sensoria
- Distributive sensation (ecstasy, intra-action, etc.)
- Sensation and temporality/time
- Nonhuman sensations and/or noncanonical sensations like thermoception or chronoception
- Nonfeeling or unfeeling

Submissions of 11,000 words or less (including endnotes and references) should be submitted electronically at www.editorialmanager.com/al/default .asp by **October 4, 2021**. For assistance with the submission process, please contact the office of *American Literature* at am-lit@duke.edu or (919) 684-3396. For inquiries about the content of the issue, please contact the coeditors: Erica Fretwell (efretwell@albany.edu) and Hsuan Hsu (hlhsu@ucdavis.edu).

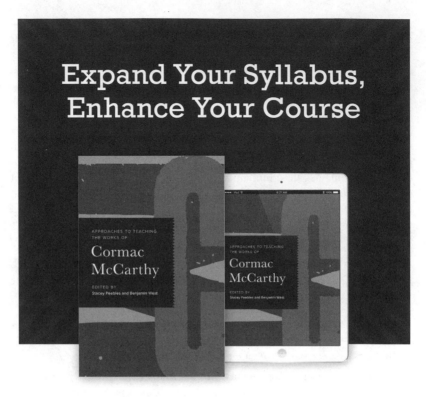